PRIMARY

D1580910

Concept, Evaluation, and Policy

Barbara Starfield, M.D., M.P.H.

Professor and Head,
Division of Health Policy
Johns Hopkins University

New York   Oxford
OXFORD UNIVERSITY PRESS
1992

Oxford University Press

Oxford   New York   Toronto
Delhi   Bombay   Calcutta   Madras   Karachi
Petaling Jaya   Singapore   Hong Kong   Tokyo
Nairobi   Dar es Salaam   Cape Town
Melbourne   Auckland
and associated companies in
Berlin   Ibadan

Copyright © 1992 by Oxford University Press, Inc.

Published by Oxford University Press, Inc.,
198 Madison Avenue, New York, New York 10016-4314
Oxford is a registered trademark of Oxford University Press

Library of Congress Cataloging-in-Publication Data

Starfield, Barbara.
Primary care : concept, evaluation, and policy /
Barbara Starfield.
p. cm.   Includes index.
ISBN 0-19-507489-0   :
ISBN 0-19-507517-X (pbk.)
1. Family medicine.   2. Family medicine—United States.
3. Medical care.   I. Title.
[DNLM: 1. Primary Health Care—organization & administration.
W84.6 S795p]   R729.5.G4S72   1992
362.1—dc20   DNLM/DLC
for Library of Congress   91-39359

3 5 7 9 8 6 4 2

Printed in the United States of America
on acid-free paper

To my family:
Robert and April, Jon, Steven and Susan, and Deborah and especially
my husband Neil (Tony) Holtzman, whose love and encouragement
has provided the basis for everything else.

PREFACE

In its most highly developed form, primary care is the point of entry into the health services system and the locus of responsibility for organizing care for patients and populations over time. There is a universally held belief that the substance of primary care is essentially simple. Nothing could be further from the truth.

This book is testimony to the challenges of primary care. It consists of three major sections:

- The concept of primary care. Two chapters trace the history of the development of primary care as a concept and lay the groundwork for assessing its contributions.
- Measurement and evaluation of the essential components of primary care. Four chapters present the importance of each major feature of primary care and approaches to assessing its attainment.
- Policy issues in primary care. This section is divided into three parts:

(a) A detailed discussion of the health system in the United States that graphically illustrates the difficulties in attaining primary care, even in a highly industrialized nation.

(b) Important issues in primary care, with chapters on primary care personnel, medical records and information systems, practitioner-patient communication, quality of primary care, and community orientation.

(c) Systems of primary care, including chapters addressing alternative ways to evaluate them; a comparison of them in different nations; a research agenda that might be adopted as a national strategy and be undertaken by research organizations or individual researchers; and a policy agenda that is based upon a synthesis of issues raised in earlier chapters.

This book is for those challenged by critical thinking about what primary care is and what it should be. This includes many clinicians who practice it, educators who teach it, researchers who try to study and evaluate it, patients who use it and want to understand it, and policymakers interested in improving it. The book has four purposes: to help practitioners of primary care to understand what they do and why, to provide a basis for the training of primary care practitioners, to stimulate research to provide a more substantive basis for improvements in primary care, and to help policymakers understand the difficulties and challenges of primary care and its importance.

This book had its genesis in a series of lectures for a course on primary care in the United States. Over the years, more and more students from other countries, including developing ones, have enrolled in the course; they say that they find the concepts and approach to measurement useful to them in thinking about their own health systems. So although the examples cited in the book are almost all from the United States and although the literature derives primarily from studies in English-language countries, the principles apply everywhere.

The reader will note that this study is oriented towards care provided by physicians. This is not to suggest that other personnel, especially nurses, have little responsibility in primary care. Rather, the book's thesis is based on the assumption, generally considered as accurate for industrialized nations, that it is the physician who bears the responsibility for the totality of primary care. Other personnel may assume responsibility for certain of its aspects, even for some highly central ones, but it is the physician who must oversee all of its aspects. Recognition of the importance of other types of professionals comes in the form of the use of the word "practitioner" throughout the book. Practitioner, instead of physician, is used wherever the particular function under consideration is sometimes assumed by a non-physician professional.

Primary care is complex; its challenges will require concerted efforts at research and systematic translation of knowledge into policy. If this book stimulates these activities, it will have achieved its purposes.

Baltimore, Maryland   B.H.S.
October, 1991

ACKNOWLEDGMENTS

First and foremost, a special thanks to my parents, Martin and Eva Starfield, who painstakingly pored over draft after draft, making needed improvements in style and readability. I owe undying affection and gratitude to them for this yeoman's effort as well as for all of their ongoing inspiration and support.

Discussions with many of my colleagues in the Department of Health Policy and Management at the Johns Hopkins University School of Hygiene and Public Health, and their critiques of my earlier writings, were invaluable in the development of ideas in this book. Several of these warrant special mention. Jonathan Weiner, my former "advisee" and now research collaborator, provided me with a wealth of background material and numerous suggestions, especially for Chapters 7 and 8. His work on primary care issues and especially his comparisons of international systems are themselves major contributions to the field. Other faculty colleagues were also helpful with their critiques and suggestions; they include Vicente Navarro, Debra Roter, Diane Rowland, and Greg DeLissovoy.

Dr. Juan Gérvas, general practitioner in Spain and primary care researcher in Europe, inspired me to dig deeper into many of the issues in this book. Our long and intensive discussions were very helpful to me in thinking through the challenges in the field, and his warm colleagueship is very much appreciated. Dr. Neil Holtzman's generosity in making available his taped encounters with physicians and patients provided many of the vignettes to illustrate the theoretical points.

Dr. Milton Roemer provided insightful comments and suggestions on the international aspects, for which I am grateful. I also owe a debt of gratitude to Dr. Michael Wolfson of Statistics, Canada, and to students at the Johns Hopkins University School of Hygiene and Public Health who provided statistics and information about the primary care characteristics of their countries. They include Ton Rijsemus, Fokke Melma, Frederick Van Loon, Petra Osinski, John Murray, and Richard Steele. Other students, especially including Christopher Forrest and Cynthia Kapphahn, critically reviewed the manuscript and made useful suggestions.

Special thanks are owed to the Agency for Health Care Policy and Research for providing funds to enable the wider distribution of this book. The agency, under the leadership of Dr. J. Jarrett Clinton, is making major inroads into the science of health care delivery in general and of primary care in particular.

Large sections of this book, and revisions to all of the chapters, were written while in residency at the Rockefeller Foundation's Study and Conference Center in Bellagio, Italy. I am deeply grateful to have been offered the opportunity to work there.

CONTENTS

ABBREVIATIONS

AAPPO	American Association of Preferred Provider Organizations
ACG	Ambulatory Care Group
AHCPR	Agency for Health Care Policy and Research
ASPN	Ambulatory Sentinel Practice Network
COBRA	Consolidated Budget Reconciliation Act
COC	Continuity of Care
COOP	Dartmouth Primary Care Cooperative Project
COPC	Community Oriented Primary Care
COSTAR	Computer Stored Ambulatory Record
DEFRA	Deficit Reduction Act
EPCI	Empirical Primary Care Index
EPO	Exclusive Provider Organization
FEC	Free-Standing Emergency Center
FFS	Fee-for Service
GMENAC	Graduate Medical Education National Advisory Committee
HCFA	Health Care Financing Administration
HIO	Health Insurance Organization
HMO	Health Maintenance Organization
ICD	International Classification of Diseases
ICH-PPC	International Classification of Health Problems in Primary Care
ICPC	International Classification of Primary Care
IOM	Institute of Medicine
IPA	Independent Practice Association
JCAHO	Joint Commission on the Accreditation of Health Care Organizations
LICON	Likelihood of Continuity
LISECON	Likelihood of Sequential Continuity
MCH	Maternal and Child Health
MIP	Managed Indemnity Plans
NAMCS	National Ambulatory Medical Care Survey
NCHS	National Center for Health Statistics
NPCI	Normative Primary Care Index
OBRA	Omnibus Reconciliation Act
PCTP	Primary Care Training Program
PHC	Primary Health Care
PHS	Public Health Service
POMR	Problem Oriented Medical Record
PPGP	Pre-Paid Group Practice
PPO	Preferred Provider Organization
PPRG	Pediatric Practice Research Group

PRO	Peer Review Organization
PSRO	Professional Standards Review Organization
S/HMO	Social HMO
SAPCI	Self-assessed Primary Care Index
SECON	Sequential Continuity
SOAP	Subjective Data; Objective Data; Assessment Data; Plan
STOR	Summary Time Oriented Record
TEFRA	Tax Equity and Fiscal Responsibility Act
TMR	The Medical Record
UPC	Usual Provider Continuity
USDHHS	U.S. Dept. of Health and Human Services
WIC	Women, Infants, and Children
WONCA	World Health Organization of National Colleges & Academies of Family Practice

I

PRIMARY CARE: CONCEPTS AND GOALS

1

What Is Primary Care?

the failure to recognize that the results of specialized observation are at best only partial truths, which require to be corrected with facts obtained by wider study. . . . No more dangerous members of our profession exist than those born into it, so to speak, as specialists.

Osler W. Remarks on specialism.
Boston Med Surg J 1892; 126; 457–459

Every health services system has two main goals. The first is to optimize the health of the population by employing the most advanced knowledge about the causation of disease, illness management, and health maximization. The second, and equally important, goal is to minimize the disparities across population subgroups to ensure equal access to health services and the ability to achieve optimal health.

As knowledge accumulates, medical professionals have become more and more specialized because of the impossibility of knowing everything about all types of health problems. Therefore, we see the tendency in every country for health professions to become more fragmented, with increasingly narrow interests and competence in treating specific types of diseases. In some countries, there are more specialists than generalists. Specialty care often commands more resources than basic care, because more technology is developed and deployed to keep seriously ill people alive than for programs to prevent illnesses or reduce discomfort from more common, non-life-threatening ailments. Although the tendency toward specialization based on the most current knowledge may provide highly efficacious care of individual diseases, it is unlikely to produce highly effective basic care. Why is this so? Specialization oriented toward treating disease cannot maximize health because preventing illness and promoting optimal functioning requires a broader perspective than can be achieved by the disease specialist. Effective medical care is not limited to the treatment of disease itself; it must consider the contexts in which the illness occurs and the patient lives. Moreover, diseases rarely exist in isolation, especially when experienced over time. Thus, disease specialists may provide the most appropriate care for the specific illnesses within their area of special competence, but a generalist is needed to integrate care for the variety of health problems that individuals experience over time.

A specialty-oriented health system has another problem: specialization threatens the goals of equity. No society has unlimited resources to provide for health services. Specialty care is more expensive than primary care and is therefore less

3

accessible to individuals with fewer resources to pay for it. Moreover, the resources needed for highly technical specialty care compete with those required to provide basic services, especially for people unable to pay for them.

Primary care is the means by which the two goals of a health services system— optimization of health and equity in distributing resources—are balanced. It is the basic level of care provided equally to everyone. It addresses the most common problems in the community by providing preventive, curative, and rehabilitative services to maximize health and well-being. It integrates care when more than one health problem exists, and deals with the context in which illness exists and influences people's responses to their health problems. It is care that organizes and rationalizes the deployment of all resources, basic as well as specialized, directed at promoting, maintaining, and improving health.

Primary care is distinguished from secondary (consultative) and tertiary (referral) care by several characteristics. Primary care deals with more common and less well-defined problems, generally in community settings such as offices, health centers, schools, or homes. Patients have direct access to an appropriate source of care, which continues over time for a variety of problems and includes needs for preventive services.

Compared with specialty medicine, primary care uses both capital and labor less intensively and is less hierarchical in organization. Therefore, it is inherently more adaptable and capable of responding to changing societal health needs. In specialty care, patients have typically been referred by another physician who has already explored the patient's problem and initiated preliminary diagnostic work. The diagnostic process results in a precise definition of pathophysiology; interventions are directed primarily at that pathophysiological process. In primary care, in contrast, the patient is usually known to the physician and usually initiates entry into the system, often with poorly specified and vague complaints. The major task is to elucidate the patient's problem and elicit information that will lead to a diagnosis and choice of the most appropriate management. Primary care physicians, in contrast to specialists, deal with a broader range of problems, in individual patients as well as across their practice population. Because they are closer to the patient's milieu than specialists, they are in a better position to appreciate social and environmental impacts on illness.

The Historical Context of Primary Care

In 1920, eight years after national health insurance was instituted in Great Britain, a white paper was released (Lord Dawson of Penn, 1920) dealing with the organization of the health services system. It distinguished three major levels of health services: primary health centers, secondary health centers, and teaching hospitals. Formal linkages among the three levels were proposed and the functions of each were described. This formulation was the basis for the concept of regionalization: a system of organization of services designed to respond to the population's various levels of need for medical services. This theoretical arrangement subsequently provided the basis for the reorganization of health services in many countries, which now have clearly defined levels of care, each with an identifiable and functioning primary medical care sector.

In 1977, at its thirtieth annual meeting, the World Health Assembly decided unanimously that the main social target of member governments should be "the attainment by all citizens of the world by the year 2000 of a level of health that will permit them to lead a socially and economically productive life". Now known as "Health for All by the Year 2000," this declaration set in motion a series of activities that have had a major impact on thinking about primary care. The principles, enunciated at a conference in Alma-Ata, dealt with the topic of "primary health care." The consensus reached there was confirmed at the World Health Assembly's subsequent meeting in May 1979. *Primary health care* (PHC) was defined as follows:

> Essential health care based on practical, scientifically sound, and socially acceptable methods and technology made universally accessible to individuals and families in the community by means acceptable to them and at a cost that the community and the country can afford to maintain at every stage of their development in a spirit of self-reliance and self-determination. It forms an integral part of both the country's health system of which it is the central function and the main focus of the overall social and economic development of the community. It is the first level of contact of individuals, the family and the community with the national health system, bringing health care as close as possible to where people live and work and constitutes the first element of a continuing health care process.   (World Health Organization, 1978)

Primary health care (PHC) was recognized as an "integral, permanent, and pervasive part of the formal health care system in all countries, and not as an 'add-on' " (Basch, 1990). The Alma Ata conference further specified that the core components of PHC were health education; environmental sanitation, especially of food and water; maternal and child health programs, including immunization and family planning; prevention of local endemic diseases; appropriate treatment of common diseases and injuries; provision of essential drugs; promotion of sound nutrition; and traditional medicine.

Although these concepts of primary health care were intended to apply to all countries, there is disagreement about the extent to which they are applicable in industrialized nations as well as about the impediments to applying them (Kaprio, 1979). The concept of PHC, in its emphasis on "nearness to the people," seems alien in countries with health systems based on technology, specialization, the primacy of the hospital, and medical school curricula under the control of hospital-based specialists. Furthermore, the principle that health care should be needs-related is not easily understood in countries with well-established health systems but no information system to document health needs systematically or evaluate the impact of health services on them. Community orientation has little historical basis in the health systems of most industrialized nations.

Furthermore, many of the specific goals defined at Alma Ata have already been achieved in industrialized countries (Vuori, 1984). Most such countries can point with pride to long-standing programs for most of the activities: food supply, safe water, maternal and child health, immunization, prevention and control of endemic diseases, basic treatment of health problems, and provision of essential drugs. When primary health care is viewed as "accessible" services, many of these countries can justly claim to have achieved the goals because of availability of medical care services. It is only when nations view PHC as a strategy to integrate all aspects of health services that it becomes equally applicable as a goal in industrialized nations. This view requires that a health care system be organized to stress social justice and equality, self-responsibility, international solidarity, and acceptance of a broad concept of health (Vuori, 1984).

Some countries are actively reorganizing their health services to consolidate the medical and health aspects of primary care. For example, the new "family doctors" in Cuba live where they work. They are therefore members of the community they serve, and it is to their advantage to act as agents of change when environmental or social circumstances need improvement (Gilpin, 1991). This integration of conventional medical services with social and environmental services fits the model envisaged at Alma Ata.

Table 1.1 describes the changes required to convert conventional primary medical care in industrialized nations to broader primary health care, as defined at Alma Ata.

All the directions of change are goals of primary care as conceived in this book, in which the term *primary care* connotes conventional primary medical care striving to achieve the goals of primary health care. The intent is to suggest that increasing the orientation of primary care services toward meeting the needs of communities as well as those of individuals who appear for care will bring conventional primary medical care closer to the vision of primary health care of Alma Ata.

TABLE 1.1.   From Primary Medical
to Primary *Health* Care

Conventional Focus	New Focus
Illness	Health
Cure	Prevention and care
Conventional Content	**New Content**
Treatment	Health promotion
Episodic care	Continuous care
Specific problems	Comprehensive care
Conventional Organization	**New Organization**
Specialists	General practitioners
Physicians	Other personnel groups
Singlehanded practice	Team
Conventional Responsibility	**New Responsibility**
Health sector alone	Intersectorial collaboration
Professional dominance	Community participation
Passive reception	Self-responsibility

Adapted from Vuori, 1984.

Describing Primary Care

Primary care, as defined earlier in this chapter, provides the philosophical underpinnings for the organization of a health services system. But the general definitions provide no help in determining whether a given system merits the description "primary care." More specificity is required.

Primary care is distinguished from other types of care by clinical characteristics of patients and their problems. These characteristics include the extent of variety of diagnoses or problems seen, an identifiable component devoted to prevention of illness, and a high proportion of patients already known to the practice.

Primary care practitioners are ordinarily distinguished from their secondary and tertiary counterparts by the variety of the problems encountered. Over the totality of problems, specialists might have a greater variety of diagnoses, since most very rare conditions might be encountered in specialty care rather than primary care. But for the most common diagnoses—those comprising 50 percent of visits—generalists would be expected to have a greater variety. In addition, by definition, they are limited to certain types of diagnoses and therefore care for a narrower range of diagnoses. Thus, primary care providers should see a greater variety of diagnoses, at least among their most frequent problems, than do specialists.

Another way to assess the variety of diagnoses or the variety of presenting problems is to examine the percentage of all diagnoses contributed by the fifty most common diagnoses; this percentage should be lower in primary care practice than in specialty practice, because of the greater variety in primary care practices.

Since primary care is the point of first contact with the health care system, its practitioners should encounter a much wider array of presenting problems than is the case in specialty care. When these presenting problems are cataloged, a much greater number of problems should comprise any given percentage of all problems in primary care than in specialty care.

Primary care practices might also be assumed to have a larger percentage of prevention-related visits.

Primary care practices should involve more patients who are continuing in care than those coming into care for the first time, and descriptions of primary care and speciality practices should reflect this difference. A related characteristic that should distinguish primary care from specialty practitioners is the greater familiarity of primary care practitioners with both the patient and the patient's problems. Both primary care physicians and specialists are expected to see new patients and "old" patients with old problems. Primary care practitioners should see more old patients with new problems, since they are responsible for the patient's care over time, regardless of the particular problem.

If primary care and specialty practices cannot be distinguished by these descriptive characteristics, there is reason to be concerned about the clear delineation of primary care within the health services system. As Chapter 7 demonstrates, such is the case in the United States.

The next chapter moves from defining and describing primary care to a framework for measuring and evaluating it. Even more specificity is needed for this than for definition and description. As the chapter shows, the most important goals of primary care can be measured in a way that provides a basis for setting goals and continuously improving their achievement.

References

Basch P. Textbook of International Health. New York, Oxford University Press, 1990, Chapter 7.

Gilpin M. Update—Cuba: On the Road to a Family Medicine Nation. J Public Health Policy 1991; 12(1):83–103.

Kaprio L. Primary Health Care in Europe. Copenhagen, Regional Office for Europe. World Health Organization, 1979.

Lord Dawson of Penn. Interim Report on the Future Provisions of Medical and Allied Services. United Kingdom Ministry of Health. Consultative Council on Medical Allied Services. London: Her Majesty's Stationery Offices, 1920.

Osler, Sir William. Remarks on specialism. Boston Med Surg J 1892; 126:457–59.

Vuori H. Primary health care in Europe—problems and solutions. Community Medicine 1984; 6:221–31.

World Health Organization. Primary Health Care. Geneva, 1978.

2

A Basis for Evaluating Primary Care

Neither definitions nor descriptions of primary care provide a basis for evaluating its attainment. All evaluations require standards against which performance can be measured, either a preset goal (the "normative" approach) or a comparison of one system (or facility) with another (the "empirical" approach) (Parker et al., 1976). This chapter provides a theoretical framework for setting these standards. The next section (chapters 3–6) shows how to use this framework to assess each of the major features of primary care.

Primary care is widely accepted as the delivery of first-contact medicine; the assumption of longitudinal responsibility for the patient, regardless of the presence or absence of disease; and the integration of physical, psychological, and social aspects of health to the limits of the health personnel's capabilities. Such a description was proposed in the Millis report (1966) and is consistent with the major features of primary care: first contact, longitudinality, comprehensiveness, and coordination (or integration) (Alpert & Charney, 1973; Parker, 1974).

One approach to assessing primary care was suggested by a committee of the Institute of Medicine (1978), which listed the attributes of primary care as accessibility, comprehensiveness, coordination, continuity, and accountability. Of these five attributes, only comprehensiveness was actually defined ("ability of the primary care team to handle problems arising in the population it serves"). Accountability was recognized as a feature not unique to primary care, although it is essential. The committee acknowledged that primary care could not be assessed by descriptive features such as the location of care or by the provider's field of training, or by the provision of a particular set of services. But it stated that "professionals who train men and women for primary care should accustom their students to a practice environment that meets or exceeds" the standards of primary care, specified in the form of positive responses to a set of twenty-one questions concerning its five attributes. Seven questions were devoted to accessibility, six to comprehensiveness, four to coordination, three to continuity, and one to accountability.

The results of the committee's efforts were an important milestone in the attempt to devise a normative method for measuring attainment of primary care. However, the checklist of questions has some limitations. First, most of the indicators in the checklist might be attributes of secondary or tertiary care as well as of primary care, including the opportunity for patients to schedule appointments; appreciation of patients' culture, background, socioeconomic status, and living circumstances; willingness to admit patients to hospital, nursing homes, or convalescent homes; provision of simple, understandable information about fees; acceptance of patients without regard to race, religion, or ethnicity; easily retrievable and accessible medical records; provision of a summary of patients' records to other physicians when needed; and assumption of responsibility for alerting proper authorities if a patient's problems reveals a health hazard that may affect others.

Second, many of the indicators may be difficult to achieve as they require a very high level of performance and allow for no variability. One example is the requirement that "90% of appropriate requests for routine appointments such as preventive examinations be met within one week."

Third, many of the indicators represent the potential ability to provide a service rather than its actual accomplishment. Examples include provision of personnel who can deal with patients with special language barriers (rather than actual provision of such services), willingness of practitioners to admit patients to other facilities (rather than doing so when it is necessary), and willingness of the practice unit to handle the great majority of patients' problems (rather than demonstrating that the unit actually accomplishes this).

An alternative to the method of assessing primary care suggested by the Institute of Medicine is measurement by assessing the actual degree of attainment rather than *potential* for achievement of first contact, longitudinality, comprehensiveness, and coordination. The standards for assessing adequacy would be based on the degree of improvement from one time to another or by comparison of one system against another rather than against an arbitrary absolute standard. As with the Institute of Medicine checklist, the approach facilitates self-evaluation of clinics or practice units as well as evaluation by an outside agency to determine the degree to which a facility or health services system provides primary care that meets accepted standards, or at least achieves a higher level of performance than others.

To do this, the criteria of the Institute of Medicine were adapted by Smith and Buesching (1986) to ascertain the degree to which primary care was achieved. They asked a random sample of people about selected characteristics of their care and derived a primary care score from the responses.

Access was ascertained from responses to statements such as "I could call my principal physician today with chest pain and get a prompt response," and "If I contacted my principal physician with a medical problem that was not an emergency, he or she would see me within a reasonable period of time."

Continuity was reflected in answers such as "My principal physician sees me for regular checkups even if I do not have a specific illness," and "My principal physician provides me with reliable follow-up treatments for illness."

Comprehensiveness was judged by such answers as "My principal physician takes care of most of my medical problems," and "My principal physician has an excellent knowledge of all my current medications."

Coordination responses included "If I have a laboratory test or x-rays, my principal physician explains the results to me," and "If several physicians are involved in my care, my principal physician organizes it."

An additional feature, personalized care, was assessed by means of responses such as "My principal physician has an excellent knowledge of the kind of work I do," and "I can discuss a personal, family, or emotional problem with my principal physician." Satisfaction with care was strongly associated with the score derived by combining the responses for these attributes; high primary care scores were also associated with patients who were ill fewer days and who stayed home fewer days because of illness, after confounding factors such as perceived health status and reported health problems were taken into account. Specialty of the principal physician was not associated with the primary care score, however, and about 25 percent of the community sample named a specialist as their principal physician.

A further refinement of the normative approach to assessing primary care employs key features of all health service systems to define each of the four attributes of primary care (*first-contact care, longitudinality, coordination,* and *comprehensiveness*) according to both the potential of the system to achieve them and the translation of this potential into the key activity. This refinement is thus distinguished from the Institute of Medicine approach, which concerns the *potential* for achieving the key feature, and from the Smith and Buesching approach, which includes only the degree of *attainment* of the key features. The next section describes these key features of all health service systems and sets the stage for selecting the elements that are important in achieving effective primary care.

Primary Care Within a Health Care System

Several factors determine the state of health of individuals and populations. Most influential is genetic structure, which has been evolving for millions of years and determines the limits of health services in improving health. The gene structure continues to evolve and tomorrow's situation, in terms of the potential for health, is likely to be very different from that of today. This is increasingly true because scientists are learning how to tamper with the gene structure to alter states of health. In addition, modern technology makes increasingly possible interference with the expression of gene structure by such means as modifying the environment, altering behavior, and using certain types of medical care.

The other determinants of health—social and physical environment, individual behaviors, and health services (medical practice) as superimposed on the genetic structure (genotype)—are shown in Figure 2.1. As indicated in the figure, the health of individuals or populations is predestined by genetic structure heavily modified by the social and physical environment, by behaviors that are culturally or socially determined, and by the nature of the health care provided. Although the health system is a less important determinant of health than the social and physical environment or behavioral characteristics, the contributions of primary care largely relate to the health services system.

Each health services system has three types of components: structural, process, and outcome (Donabedian, 1966). Figure 2.2 shows these components. The individual characteristics within each component differ from place to place and from time to time, but each health services system has a structure consisting of the characteristics that enable it to provide services, the processes involving actions by

GENOTYPE

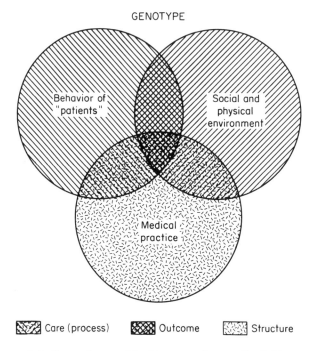

FIGURE 2.1. Determinants of health status. *Source:* Starfield, 1973.

the practitioners in the system as well as the actions of the populations and patients, and the outcome as reflected in various aspects of health status. It should be remembered that, as Figure 2.1 shows, the resulting outcome of a health system is greatly influenced by individual behavior and the environment as well as by the health system itself.

The following pages describe an approach to measuring primary care that is based on certain structures and processes within the health services system. This approach to measuring primary care assumes these structural attributes must be in place for important activities to occur. It also assumes the importance of assessing the performance of those activities. Thus, measuring the key features involves measurement of a behavioral characteristic and the structural characteristic on which it depends.

Let us examine, in turn, the structure, the processes, and the outcome of the health services system, to set the stage for choosing those that are most important in primary care.

Structure of a Health Services System

Structure consists of the resources needed to provide services. As Figure 2.2 shows, there are nine main structural components:

Personnel involved in providing the services, their education and their training.

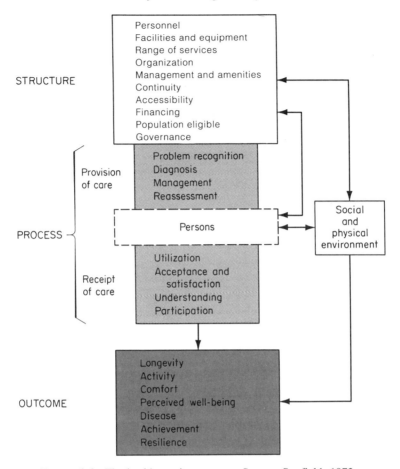

FIGURE 2.2. The health services system. *Source:* Starfield, 1973.

Facilities and equipment: the buildings, including the hospitals, clinics or health centers, and offices, as well as the physical components of the facilities, including such elements as laboratory instruments and technology for diagnosis or therapy.

Management and amenities, including characteristics of services other than those directly related to clinical care. For example, are laboratory results reported in a timely fashion? Are patients treated with courtesy and respect?

Range of services provided by facilities. This range of services may vary from country to country and from community to community, but every facility has made decisions about what kinds of services will be available and what will not be available. The range of services provided is an important consideration for the nature of primary care and is discussed in greater detail in the chapter on comprehensiveness.

Organization of services. Do the personnel work in groups or alone? What are the mechanisms for ensuring accountability, and who is responsible for providing the different aspects of care?

Mechanisms for providing continuity of care. These mechanisms are especially important in primary care because without them there would be no way to deal with problems that require more than one visit or the transfer of information. Continuity is usually provided in the form of practitioners or teams of practitioners who serve as the primary contact for the patient, but sometimes the only mechanism for continuity is some form of written record. Continuity is considered in greater detail in the chapter on coordination.

Mechanisms for providing access to care. There is no point to having personnel, facilities, and equipment if they cannot be reached by persons who need them. There are several types of accessibility: accessibility in time (that is, the hours of availability), geographic accessibility (adequacy of transportation and distance to be covered), and psychosocial accessibility (Are there language or cultural barriers to communication between personnel in the facilities and patients?). Accessibility and its special importance in primary care is considered in greater detail in the chapter on first contact care.

Arrangements for financing. What is the method of payment for services and how are the personnel remunerated for their work? Of all of the structural features, this is the most likely to differ across countries and is therefore of great interest for cross-national comparative studies.

Delineation of the population eligible to receive services. Each unit of the health services system should be able to define the community it serves and should know its important sociodemographic and health characteristics. Members of the population should be able to identify their source of care and be aware of its responsibility for providing required services. This structural feature is another critical element for primary care, especially for the feature known as *longitudinality*. It is discussed further in a chapter devoted to this subject.

Governance of the health system. Health systems differ in their accountability to those they serve. Often they do not involve the population in decisions about the way services are organized or delivered. Sometimes, community councils serve in an advisory capacity. Rarely, responsibility for decision-making is shared or assumed by community boards.

Processes of a Health Services System

The processes of a health services system have two components: those that represent activities of the providers of care and those that represent activities of the population. The providers must first recognize the needs existing in both the community and individual patients. This feature is known as *problem (or needs) recognition* and is a particularly important consideration for primary care. The problem may be a symptom, a sign, an abnormal laboratory test, a previous but relevant item in the history of the patient or the community, or a need for an indicated preventive procedure. Problem recognition implies awareness of the existence of situations requiring attention in a health context. After recognizing the problem, the health professional generally formulates a *diagnosis*. This is necessary to move to the next step in the process of care, to institute an appropriate strategy for *treatment* or *management*. Subsequently, from the viewpoint of the professional's activities, arrangement must be made for *reassessment* of the problem to determine if the original recognition of the problem, the diagnosis, and the therapy were adequate. At this point, the process of care is started on a new cycle of monitoring and surveillance, with recognition of the problems as they now exist.

The processes of care that reflect how people interact with the health system are also important. First, people decide whether and when to use the health care system. If they do use it, they come to an understanding of what providers offer them, and then decide how satisfied they are with their care and whether or not they will accept the providers' recommendations or instructions. Subsequently, they decide on the extent to which they participate in the process. They can decide to carry through the recommendations, to modify them in ways they see fit, or to disregard them partly or completely. Certain processes of care contributed by patients are a key consideration in the assessment of primary care, as is noted in subsequent chapters on first contact and longitudinality.

Outcome of Care in the Health Services System

The structure of the health system and its translation into the processes of care has an impact on health status, an effect known as *outcome of care*. There are many ways to consider outcome of care (McDowell and Newell, 1987); Figure 2.3 presents one method. In this conceptualization, health status has seven components, ranging from longevity to resilience.

Longevity: The most common measure of health status, especially at the population level, is longevity, or life expectancy, and its converse, mortality. An important characteristic of the health of individuals is their life expectancy; the average life

LONGEVITY	Normal Life Expectancy			Dead
ACTIVITY	Functional			Disabled
COMFORT	Comfortable			Uncomfortable
PERCIEVED WELL-BEING	Satisfied			Dissatisfied
DISEASE	Not Detectable	Asymptomatic	Temporary	Permanent
ACHIEVEMENT	Achieving			Not Achieving
RESILIENCE	Resilient			Vulnerable

FIGURE 2.3. Health status/outcome of care. *Source:* Starfield, 1974.

expectancy in a population is an important descriptor of the health status of a nation. Health care systems influence life expectancy, but it is also affected by such other determinants as genetic structure, the social and physical environment, and personal behaviors.

Activity: The second component of health status is the nature of activity of the individual or population. Relevant qualities include those pertaining to the kinds of disability that affect the individual and, on the population level, the proportion of the population that can carry on with normal activities.

Comfort: This includes pain or other sensations that interfere with work and pleasure.

Perceived well-being and satisfaction: This characteristic connotes how people view their own health and the extent to which they are satisfied with it.

Disease: This involves the presence of conditions recognized as potentially or actually interfering with the well-being of individuals or the population; it includes mental as well as physical pathology.

Achievement: This reflects the positive aspects of health that must be considered in achieving what the World Health Organization has defined as "a state of well-being." Achievement signifies the level of development or accomplishment and the potential for future development of better health. A common way of describing achievement concerns how normal social roles are performed.

Resilience: This characteristic of health also pertains to a state of well-being. This category refers to the ability to cope with adversity and measures the potential for resisting a range of possible threats to health. Ability to respond constructively to stress may be measured by physiological techniques, psychological techniques, or evidence that certain defenses known to increase resistance are present or have been provided. The prototype of biologic resilience is the state of being appropriately immunized against preventable diseases. A second measure of resilience is the attainment of certain nutritional standards. A third measure is the performance of certain health behaviors known to reduce the likelihood of disease; a typical example is a definable level of physical exercise.

In sum, an assessment of the effectiveness of a health system, whether it concerns the individual, the community, or the entire population, should take into account at least some aspects of *all* of these features of health status. To achieve equity among population subgroups, it is also necessary to have information about their health status so as to determine whether the health of vulnerable population groups differs from that of the rest of the population and where such differences exist.

Measuring the Attainment of Primary Care: The Structure–Process–Outcome Approach

Both the potential for and the achievement of the critical features of primary care can be measured by the *structure–process–outcome* approach. Four structural elements of the health services system define the potential, and two process features translate the potential into the important activity.

The four structural elements—accessibility, range of services, eligible population, and continuity—are similar to the indicators suggested in the Institute of Medicine report. They are defined as follows:

Accessibility involves the location of the facility near the population it serves, the hours and days it is open for care, the degree to which it can tolerate visits made without appointments, and the extent to which the population perceives these aspects of accessibility as convenient.

Range of services is the package of services available to the population, as well as those services the population believes are available.

Definition of the eligible population includes the degree to which a health care service can identify the population for which it assumes responsibility and to which the individuals in the served population know they are considered part of it.

Continuity consists of the arrangements by which care is provided as an uninterrupted succession of events. Continuity may be achieved by a variety of mechanisms: one practitioner who cares for the patient, a medical record that reflects the care given, a computer record, or even a client-held record. The extent to which the facility provides such arrangements and the perception of their attainment by the individuals in the population define the extent of continuity of care.

Translating potential into appropriate activity requires two process elements of the health services system: utilization of services by the population and recognition of problems by health services practitioners.

Utilization refers to the extent and kind of use of health services. The primary reason for a visit may be to investigate the occurrence of a new problem, to follow up an old one, or to receive preventive services. Utilization may be initiated by the patient or be at the request or direction of a health professional, or it may occur as the result of some administrative requirement.

Problem (or needs) recognition is the step that precedes the diagnostic process. If problems or health needs are not recognized, there will be no diagnostic process or an inappropriate one. Patients may not complain of problems because they are not aware of them, or they may complain of one thing when the real problem is another. Accurate determination of the patient's or population's health needs is the role of the practitioner.

As previously stated, one of the four elements of structure (accessibility, range of services, eligible population, continuity) and one of the two process elements (utilization, problem recognition) are required to measure the potential for and attainment of each of the attributes of primary care: first contact care, longitudinality, comprehensiveness, and coordination. Although each of these four attributes is of sufficient importance to warrant a separate chapter, the following briefly describes the elements of structure and process required to measure each.

First contact care implies accessibility to and use of services for each new problem or new episode of a problem for which people seek health care. Regardless of what a facility states or perceives its accessibility to be, it does not provide first contact care unless its potential users perceive it to be accessible and reflect this in their use. Therefore, measurement of first contact care involves evaluating accessibility (the structural element) and utilization (the process element).

Several important questions concern first contact care. To what extent does the system provide for easy access, both geographically and with longer hours of availability? Does the population perceive access to be convenient? To what extent is easier access associated with utilization of the facility by its defined population for new problems?

First contact care is more fully discussed in Chapter 3.

Longitudinality presupposes the existence of a regular source of care and its use over time. Thus, the primary care unit must be able to identify its eligible population, and the individuals in that population, who should obtain care from the unit except when outside consultation and referral are required.

Several important questions concern longitudinality. Do individuals who are clearly identified as enrollees identify the facility as their regular source of care and use it as such over a period of time? Do all visits except those initiated by providers take place at the facility?

Longitudinality is the subject of Chapter 4.

Comprehensiveness implies that primary care facilities must be able to arrange for all types of health care services, even those not provided efficiently within the facility. This includes referrals to secondary services for consultation, tertiary services for definitive management of specific conditions, and essential supporting services such as home care and other community services. Although each primary care facility may define its range of services differently, each should make its responsibility explicit to both its patient population and the staff and must recognize situations in which services are available. The staff should provide for and recognize the need for preventive services as well as for dealing with symptoms, signs, and diagnoses of manifest illness. It should also adequately recognize problems of all types, be they functional, organic, or social. The latter is particularly important because all health problems occur within a social setting that often predisposes to or causes disease.

Several important questions concern comprehensiveness. How inclusive is the benefit package offered? Is it explicit and is it understood by the population? In providing services, do the practitioners recognize a broad spectrum of needs within the population?

Comprehensiveness is the subject of Chapter 5.

Coordination (integration) of care requires some form of continuity, either by practitioners, medical records, or both, as well as problem recognition (a process element). For example, the status of problems noted in previous visits or problems for which referrals to other practitioners were made should be ascertained at subsequent visits. This recognition of problems will be facilitated if the same practitioner sees the patient on follow-up or if there is a medical record that highlights these problems. Thus, both problem recognition and continuity are necessary in assessing coordination of care.

Several important questions concern coordination. To what extent is scheduling arranged to allow patients to see the same provider for all visits? Do the medical records contain information pertinent to the care of patients? Is there an increased recognition of problems associated with improved continuity? Is this increased recognition a function of better records, of regular practitioner continuity, or both?

Coordination is the subject of Chapter 6.

Assessing the Effectiveness of Primary Care Systems

Policymaking groups may set standards for first contact, longitudinality, comprehensiveness, and coordination and decide which services qualify for the designation of primary care. Subsequently, services that qualify as primary care can be compared with other forms of care with regard to their impact of the services. Several questions might be asked about impact. Is the attainment of a satisfactory level of first contact care associated with increased satisfaction among patients as well as with better problem resolution? Is longitudinality associated with better problem recognition and better understanding and participation of patients, and does it result in fewer days of disability and discomfort? Is comprehensiveness associated with different utilization patterns, fewer episodes of new illness, or more rapid resolution of problems? Is coordination associated with less overall utilization, better understanding and increased patient participation in their care, and more rapid problem resolution with fewer new problems? Does primary care better return patients to optimum levels of activity, comfort, and satisfaction with their health, and does it do a better job helping people achieve their full potential and maximal resilience against threats to their health? That is, does primary care produce better health status, as defined by the characteristics in Figure 2.3? Are patients better served by a health system formally organized into primary care, secondary care, and tertiary care than they are by a system permitting patients to chose the type of practitioner each time they perceive a need for care?

No primary care system can attain perfect performance in all four essential components (first contact, longitudinality, comprehensiveness, and coordination). If standards are too high, patients will be disappointed and professionals will be frustrated. But justification for primary care need not depend on attaining optimum standards; it is sufficient to demonstrate that the goals of primary care are better served by practitioners trained and organized to provide primary care than by practitioners trained to focus on particular illnesses, organ systems, or pathogenetic mechanisms and that the attainment of goals improves progressively over time. The following section is devoted to this theme. Its four chapters deal with the measurement of the four unique characteristics of primary care. All chapters are organized in the same way.

- A brief introduction, which defines the characteristic and states why it is important.
- A summary of what is known about the benefits of the characteristic.
- A section concerning the policy and research implications of the characteristic.
- Alternative methods for measuring the structural element of the characteristic, first from the vantage of the population and then from the vantage of the facility providing

care are presented. This structural element provides the potential for achieving the characteristic.

- Alternative methods for measuring the process element of the characteristic, first from the population vantage and then from the vantage of the facility are presented. This process element is the behavioral component of the characteristic.
- A summary of what is known about attaining the characteristic (both its structural and behavioral elements).

In theory, the structural element of each characteristic should be closely related to the process element. That is, better access should lead to better utilization for each new health problem; better identification with a regular source of care should be associated with more consistent use of that regular source of care over time; a broader range of available services should be associated with better recognition of the need for those services; and better continuity of care should lead to better coordination of care as measured by greater recognition of information about patients. In practice, little research specifically tests the theory.

Furthermore, the unique features of primary care are not always clearly separable. If there is a relationship between a practitioner and patient that transcends the presence of particular problems or types of problems (longitudinality), it is more likely that the practitioner will be the one from whom care is initially sought for a new problem (first contact care). Similarly, longitudinality should also be related to comprehensiveness, as a practitioner or facility that provides care over time regardless of the type of problem should also be providing a greater breadth of services. Similarly, the greater the range of services that are provided, the greater the burden of coordination, especially if some of these services have to be provided somewhere other than the primary care site itself. Despite these interrelationships, the four unique characteristics of primary care are conceptually distinct; only when they are put into practice does the potential for overlap become evident. The extent of overlap is, in fact, a high priority for research.

The research literature is generally sparse in the are of primary care. Especially lacking are studies of the impact of primary care or its components on health status. The absence of a literature linking characteristics of the health services system to outcomes is hardly limited to primary care, however. As with health services research in general, this remains an important agenda for the future (Bowen, 1987; Roper et al., 1988).

References

Alpert J and Charney E. The Education of Physicians for Primary Care. Publication (HRA) 74-3113. Rockville, Md.: U.S. Department of Health, Education, and Welfare, Public Health Service, Health Resources Administration, 1973.

Bowen O. Shattuck lecture—What is quality care? N Engl J Med 1987; 316:1578–80.

Donabedian A. Evaluating the quality of medical care. Milbank Mem Fund Q 1966; 44(part 2):166–206.

Institute of Medicine. A Manpower Policy for Primary Health Care. IOM Publication 78–02. Washington, D.C.: National Academy of Sciences, 1978.

McDowell I and Newell C. Measuring Health. A Guide to Rating Scales and Questionnaires. New York: Oxford University Press, 1987.

Millis JS (Chairman). The Graduate Education of Physicians. Report of the Citizens Commission on Graduate Medical Education. Chicago: American Medical Association, 1966, p. 37.

Parker A, Walsh J, and Coon M. A normative approach to the definition of primary health care. Milbank Mem Fund Q 1976; 54:415–38.

Parker A. The Dimensions of Primary Care: Blueprints for Change. In Primary Care: Where Medicine Fails. S. Andreopoulos (ed.). New York: John Wiley & Sons, 1974.

Roper W, Winkenwerder W, Hackbarth G, and Krakauer H. Effectiveness in Health Care: An initiative to evaluate and improve medical practice. N Engl J Med 1988; 319:1197–202.

Smith W and Buesching D. Measures of primary medical care and patient characteristics. J Amb Care Management 1986; 9:49–57.

Starfield B. Health Services Research: A working model. N Engl J Med 1973; 289:132–36.

Starfield B. Measurement of outcome: a proposed scheme. Milbank Mem Fund Q 1974; 52:39–50.

II

ELEMENTS OF PRIMARY CARE

3

First Contact Care and Gatekeepers

When I was a resident, I had an experience that changed my career. I started in surgery and often patients would come to the clinic with aches and pains and they didn't have any surgical problem, so I would refer them to the appropriate clinic and give them the phone number to call. I assumed that I'd taken care of the problem. But one day I called the phone number just to see what patients were experiencing. The first time I called I got cut off. So I tried again and I got cut off again, and the next time I was put on hold and the next time they told me they were going to transfer the call. It ended up taking me 14 phone calls to make an appointment. And I speak English. So I started worrying about how I was referring my patients, where they are going and what we are doing to them. We function ok because we know the system, but most folks go home and they run into barriers right away and don't know how to handle them. I switched from surgery to family medicine and now I do as much for them as I can. Only when I feel I can't do any more do I refer them and I don't let them go until the appointment is made. I'm available to take care of diseases but that's not all I do. My patients are much larger than their diseases. Now I feel that I am really doing something for patients—that there's really something to practicing medicine. The best thing we can do for patients is to understand what they say, and a prerequisite is listening. Obviously I take care of their diseases—I'm available to do that in ways that they need me to do that—but that's a small part of what I do.

Dr. C, a physician

Inherent in the organization of health services by level of care (primary, secondary, tertiary) is the notion that there is a point of entry each time care is needed for a new health problem. In current parlance, this point of first contact is known as the *gatekeeper*.

Gatekeepers make sense. First, most people do not know enough about the technical details of medical care to make informed judgments about the appropriate source of care for many of their problems. As a result, adequate care may not be obtained or may be delayed and additional expense incurred. Advice and guidance from a primary care physician could be expected to facilitate the selection of the best source of care. Second, care from certain types of practitioners, particularly specialists, is more costly than care from generalists. Therefore, bypassing the generalist to go directly to the specialist is excessively costly when the primary care physician could have managed the problem less expensively.

Not every physician is equally skilled in all facets of medical care, and the entry level of care provides special challenges. Problems brought by patients are often vague and unrelated to particular organ systems. Because of varying thresholds for seeking care, some patients seek care earlier in the course of their disease then others. Therefore, primary care physicians confront a much greater variety of presentations of illness than do specialists, who generally see patients in later and more differentiated stages of illness. Primary care physicians work in the community, where the manifestations of illness are heavily influenced by their social context. In contrast, specialists, especially those who work in hospitals, deal with problems more removed from their social context.

Evidence for the benefits of first contact is important in supporting its inclusion as a key characteristic of primary care. The following section considers these benefits.

Benefits of First Contact Care

Does initial seeking of care from a primary care source result in better or less costly care? What role can a gatekeeper play?

There is little question that facilitation of access to care in general is associated with lower mortality and morbidity. The nature of the evidence has been summarized for child health (Starfield, 1985); similar types of evidence are available for adults. For example, because of budget deficits, many individuals were no longer permitted to receive outpatient services from a Veterans Administration Hospital that they had previously used as a source of care. When questioned, they were much less likely to report that they had access to needed medical care and that they actually had been receiving care. They also were more likely to report that they were in poor health, and were more likely to have poor blood pressure control one year later than those kept in treatment, even though the latter were initially more sick (Fihn and Wicher, 1988). When the state of California terminated health care benefits for many indigent adults, there was a decline in the use of services and a deterioration of health status that persisted for at least a year (Lurie et al., 1986).

Maximization of access to a *primary care source* is also important. Several studies indicate that patients should see a generalist before consulting a specialist.

One experiment required those enrolled in a health insurance program to receive care first from a generalist physician; no hospitalizations or visits to a specialist were allowed without a referral from the primary care physician. After a year of enrollment, those in the program spent fewer days in the hospital than those whose insurance permitted them to seek initial care anywhere (Moore, 1979). However, a subsequent evaluation failed to show any differences in costs of either inpatient or outpatient services, largely because the primary care gatekeeper was not able to control the practices of specialists, who were paid on a fee-for-service basis with no financial risk. In fact, 70 percent of the hospitalizations were controlled by specialists rather than by the primary care physician (Moore et al., 1983).

Studies in Canada showed that appropriate indications for tonsillectomy and/or adenoidectomy were more often present in children who had been referred from a pediatrician or who had some pediatric contact than in children who had visited only an ear, nose, and throat specialist. The outcomes of care were also better for the children seen initially by a pediatrician; such children had fewer postoperative complications and greater decreases in respiratory episodes and in episodes of otitis media following surgery (Roos, 1979).

In a study comparing visit rates in two prepaid group practices (Starfield, 1983), children in the group practice plan that required a referral from a primary care provider before visits to specialists had fewer visits to specialists than children in the plan not requiring referral. There is no evidence that the additional visits in the latter plan led to better health of the children. Because services of specialists are more expensive than those of primary care physicians, the costs of the care for children whose parents took them directly to specialists were almost certainly higher in that plan.

When a visit to a gatekeeper is required before a visit can be made elsewhere, utilization of both speciality services and emergency room visits is reduced. In a study in which new enrollees were randomized either to a system that required visits to a gatekeeper or to one with equal benefits but no gatekeeper requirement, patients in the latter plan had an average of 0.3 fewer visits to a specialist over a one-year period than patients in the other group (Martin et al., 1989). In another study of stratified random samples of patients enrolled in four Medicaid demonstration programs that required a gatekeeper, patients had large reductions in the proportion of patients with at least one emergency room visit, ranging from 27 to 37 percent for children and 30 to 45 percent for adults (Hurley et al., 1989).

Patients in multispecialty group practices with more than the mean percentage (54 percent) of primary care physicians, or in groups in which family physicians comprise more than 20 percent of the physicians, appear to spend fewer days per year in the hospital. However, the findings require confirmation since they derive from a preliminary study that did not control for other characteristics that influence hospitalization (Catlin et al., 1983).

Policy and Research Implications of First Contact Care: The Gatekeeper

Even in the United States, the notion of a gatekeeper is not new. From 1953 to 1978 patients were not permitted to see specialists at the Hunterdon Medical Center in New Jersey unless referred by primary care physicians. The gatekeeper is also often used in the military to control access to specialists. By the late 1980s, over 85 percent of health maintenance organizations restricted direct consultation of specialists by patients, and over half required the patient to visit the primary care provider before they could see a specialist (Reagan, 1987).

In systems which traditionally had free choice of physician and unimpeded access by patients to specialists, the benefit of a gatekeeper may not be intuitively obvious. If people believe that a specialist has greater expertise and skill, and they believe they are competent to decide on an appropriate one for their problem, they will feel deprived of the "best" care if they have to seek care elsewhere first. If they are correct in their assessment of their problem, unimpeded access to a specialist would save the cost of an intermediary visit. Unimpeded access is also more convenient for patients, who only have to make time for one visit, and it could also get patients under care more quickly.

On the other hand, the training of primary care physicians makes them more familiar with early stages of illness, and hence they are better able to assess the relative importance of various symptoms and signs at these stages. Their knowledge of patients makes it easier for them to evaluate the nature of changing symptoms and signs, and they are therefore likely to be more efficient in their use of resources to assess the significance of these changes. Primary care physicians might also facilitate access to specialists and direct the patient to a more appropriate one; these might shorten the time to appropriate care rather than lengthen it.

Thus, the issue of gatekeepers is controversial, and is becoming increasingly so as more types of health care organizations adopt it. There is widespread suspicion that in some plans the basis for gatekeepers is cost-containment rather than rationality of organization. Gatekeepers are increasingly being used to deter utilization of specialists, thus raising the possibility that needed care will be denied. Failure of the gatekeeper to refer might even lead to inappropriate care if the gatekeeper is not sufficiently knowledgeable or skilled in the diagnosis or management of the problem. The requirement to first seek care from a gatekeeper could delay necessary care and might lead to more utilization and hence higher costs than a direct visit to a specialist. A second issue concerns the potential restriction of free choice of provider. When the client enrolls in the health system, he or she may choose a physician (at least from a panel of physicians), but each subsequent contact must be through this physician. Choice of specialist might even be maintained at the point of each referral, but this choice would likely be restricted to a relatively small group of specialists. There are major concerns about equity with the imposition of gatekeepers when responsibility for payment lies with governmental programs or with employers. Patients who rely on these agencies for payment for services are at special risk of control over their use of services.

Other ethical issues arise in the formalization of the gatekeeper concept. When restriction of access to specialists is linked to financial incentives for the primary care physician, there is a potential conflict of interest between physicians' concerns about their income and concern about the welfare of patients. The right of patients to know about potential conflicts needs protection, and maybe formal disclosure of gatekeeper arrangements, perhaps with written consent of patients and physicians, should be routine upon enrollment in health care plans (Levinson, 1987). Conditions where restricted access could be relaxed might also be explicitly specified.

Most of these considerations concerning gatekeepers do not arise in countries where the concept of a point of first contact has a long tradition. What makes the situation different in the United States is its linkage with a cost-containment imperative, which raises the specter of rationing rather than rational organization.

If the purpose of gatekeepers is to provide more rational use of resources, there should be a scientific rationale for the belief that primary care practitioners can efficiently and effectively judge who should be referred to a specialist and who should not. Without a clear rationale for what should be handled in primary care and what should be referred, there will always be suspicion that factors such as cost-savings are controlling the decisions.

Rates of referral from generalists to specialists vary considerably, although, on average, about 5 percent of visits in general practice result in referral to a specialist (Christensen et al., 1989; Wilkin and Smith, 1987). Some variation results from differences in patients' characteristics, especially age (Penchansky and Fox, 1970), and some is associated with types of organization. For example, physicians in prepaid group practices refer less than physicians in other types of practices, and family physicians refer less than internists (Perkoff, 1978). However, variability remains even after these characteristics are taken into account. For example, Penchansky and Fox (1970) found a degree of variation from 2 to 18 percent of patients referred among internists and from 2 to 10 percent of patients referred among pediatricians. There is also considerable variability in referrals by general practitioners in both the United States and in England (Wilkin and Dornan, 1990).

One important determinant of referral rates is the availability of specialists in the community. For example, the number of outpatients seen in the specialty clinics in the different regions of Great Britain is strongly associated with the number of consultants and only weakly associated with illness rates, as measured by standardized mortality ratios and by the mean number of prescriptions per patient written by general practitioners (Roland and Morris, 1988). In Denmark, the rate of referrals among the 141 general practitioners in an entire county was most highly related to the number of specialists in different areas within the county (Christensen et al., 1989).

In the United States, where patients generally have free access to specialists without a referral from a primary care physician, the rate at which patients are seen by specialists is directly proportional to the ability of patients to go or return to a specialist without the advice or guidance of a primary care physician (Perkoff, 1978; Starfield, 1983).

Unfortunately, little information provides a basis for deciding just what needs referral and what can be appropriately diagnosed and managed by the primary care physician. Some types of referral are short term and largely for the purpose of obtaining advice, guidance, or a second opinion. Lawrence and Dorsey (1976) categorize such reasons for referral as follows:

- A need for specialized diagnostic or treatment procedure that the primary care physician does not provide. These types of referral are characterized by a relatively clear perception by the patient, the primary care physician, and the specialty physician as to what is to be done; the duration of the referral is brief and the patient never leaves the area of the primary care physician.
- Reassurance to the patient or physician regarding a second opinion when the patient or a third party payor has some doubt about a suggested course of action, there is reason to suspect that criteria other than medical justifiability might be operating, such as financial incentives or threats of malpractice, or there is legitimate uncertainty about a proposed course of action for a medical problem.
- For more extensive or more general evaluation, as in situations in which the patient's problem is complex and unclear. In these situations, the consultant is more likely to take greater initiative and the period of consultation may be somewhat longer than for the two previously mentioned reasons for referral.

In each of these situations the referral is for purposes of consultation, the period of referral is relatively brief and there is a clear expectation that the patient will return to the primary physician within a generally prespecified period of time.

In contrast, some types of referrals are relatively long-term and may involve significant disruption or even abandonment of their roles by primary care physicians. The reasons for this type of referral are of two types:

1. For unusual or serious chronic conditions for which management requires technical knowledge present only in major medical centers where ongoing research provides the most current information on new therapies and interventions.
2. For logistic reasons, convenience, or feasibility, such as the availability of financial resources that pay for services only if provided in particular types of facilities. This category of referrals should be minimized if relative predictability is desirable in a system.

The indications for referral for individual conditions are included in only a few textbooks (for example, Dershewitz, 1988; Gorell et al., 1981). In these textbooks, the indications specified are almost all related to the presence of certain symptoms or signs rather than diagnoses, suggesting that more referrals are for diagnostic or therapeutic advice rather than for ongoing care. Several other types of indications also suggest a need for short-term consultation rather than for long-term referral. These include confirmation of diagnosis; elucidation of etiology; routine assessment requiring specialized skills such as monitoring, social services, and supportive therapy; and medication titrating. Another type of indication, refractory or recurrent problems, might require either long-term referral or short-term consultation, depending on the problem and its solution.

Very little is known about the relative frequency of referrals to secondary and tertiary care. One study of 1,014 encounters reported by forty family physicians provided some information on the issue; 90 percent of encounters were managed in the primary care setting, 7 percent were categorized as "shared" care in which only a consultation was requested, and 3 percent were referrals for "supportive" (referral) care (Taylor, 1981). But this was a relatively small study of an unrepresentative group of primary care physicians.

The increasing attention to and use of gatekeepers compels attention to the nature of specialty care as well as to primary care. Research will provide planners with the basis for formalizing criteria for referrals to reduce reasons other than medical need. Specification of justifiable criteria for referral will also facilitate the development of systems to enhance the ability of primary care physicians to coordinate care—a major corollary of a rational gatekeeper role. The issues involved in improving coordination of care are pursued in Chapter 6.

Measurement of First Contact Care

Inherent in the concept of *first contact* is the notion that there should be one particular place or health care provider serving as a point of entry into the health system each time a new problem is experienced. Despite misgivings about the widespread adoption of *gatekeepers,* the concept of a place or person of first contact has become well accepted as a desirable approach to organizing services. Moreover, there is general agreement that the gatekeeper should be a generalist physician, usually a family/general practitioner, a general internist, or a pediatrician. This source should be accessible to the population who would use it whenever new problems arise.

The following section reviews ways of measuring the structural component of first contact care—that is, accessibility. The subsequent section addresses measurement of the process (behavioral) component—that is, utilization.

Measuring Accessibility

Access is the "structural" element necessary for first care. The place of care must be easily accessible and available; if not, care will be delayed, perhaps to the point of adversely affecting the diagnosis and management of the problem.

There are various types of access. Donabedian (1973) distinguishes socio-organizational access from geographic access. The former includes those characteristics of resources that either facilitate or hinder people's efforts to reach care. An example is the requirement that patients pay a visit fee before receiving service, which may provide a barrier to access. Another is less explicit social prejudices such as age, race, or social class. Geographic access, on the other hand, involves characteristics related to distance and time required to reach and obtain the services.

Access can be measured at both the systems (population) level and the facilities (service) level.

At the *population level,* accessibility can be measured by a variety of approaches.

Surveys of barriers to the use of services. National health surveys often ascertain information about the existence of various impediments to the use of services (such as the absence of insurance or other third-party coverage of the costs of services) as well as information on the extent to which needed services are not received and the reasons why. The following types of questions were asked at various times:

- Do you have to take time off from work to go to the doctor?
- About how long does it take for you to get to the doctor?
- Does the doctor speak you native language (Edwards and Berlin, 1989)?

Several other surveys in the United States contain questions on access to health services (Kovar, 1989).

Surveys of people's perceptions about accessibility of care. Penchansky and Thomas (1981) described a survey in which people were asked sixteen questions about accessibility to care. Access was divided into five types: availability, accessibility, accommodation, affordability, and acceptability. The questions were as follows:

Availability:

- All things considered, how much confidence do you have in being able to get good medical care for you and your family when you need it?

- How satisfied are you with your ability to find one good doctor to treat the whole family?
- How satisfied are you with your knowledge of where to get health care?
- How satisfied are you with your ability to get medical care in an emergency?

Accessibility:

- How satisfied are you with how convenient your physician's offices are to your home? How difficult is it for you to get to your physician's office?

Accommodation:

- How satisfied are you with how long you have to wait to get an appointment? How satisfied are you with how convenient your physician's office hours are? How satisfied are you with how long you have to wait in the waiting room? How satisfied are you with how easy it is to get in touch with your physician?

Affordability:

- How satisfied are you with your health insurance?
- How satisfied are you with the doctor's prices?
- How satisfied are you with how soon you need to pay the doctor's bill?

Acceptability:

- How satisfied are you with the appearance of the doctors offices?
- How satisfied are you with the neighborhoods their offices are in?
- How satisfied are you with the other patients you usually see at the doctor's offices?

To the extent that people's thresholds for adequacy are similar, responses to surveys such as these could provide a basis for judging the adequacy of accessibility or for comparing its adequacy in different areas. If, however, different groups of people have different standards, such a survey may conclude no differences exist when, in fact, they are considerable.

The Use/Disability Ratio. This measure of accessibility (Aday et al., 1980), like the previous one, is applied to population surveys. Respondents are asked whether there were days in which they were unable to perform their normal activities because of a health problem and whether they sought medical care. The lower the use/disability ratio, the lower the access to care. Obviously, the measure is useful only for comparative purposes, since there is no "correct" answer. Groups of the population that differ in their average use/disability ratio are presumed to differ in their access to care. The major problems with the measure are that both the numerator and the denominator depend on recall by the respondent and that it assumes that all disabilities require similar medical care or that different populations have similar disabilities. Both assumptions may be incorrect. Another problem with the use of the use/disability ratio is that the number of physician visits used in the numerator may not be linked to the number of disabilities in the denominator (Yergan et al., 1981).

The Symptoms/Response Ratio. This measure, also used in population surveys

(Andersen, 1978), elicits information about respondents' particular symptoms that are generally judged by professionals to require medical care. A panel of physicians provided the proportion of patients that should have sought care for the symptoms. The "response" is whether or not they actually sought medical care. The ratio is derived as follows:

$$\frac{\text{(persons contacting)} - \text{(those needing contact)}}{\text{those needing contact}}$$

As with the use/disability ratio, the measure is useful only for comparative purposes; population groups with different ratios presumably differ in their access to care. The major limitation of the measure is that the judgments of physicians regarding the need for care are based on opinion rather than on data. Also, the measure elicits a low frequency of symptom complexes (Yergan et al., 1981), so that large samples are required for stable estimates. Another problem is that all symptom complexes are counted rather than just one per patient.

The Episode of Illness Analysis. As originally used (Richardson, 1970) this measure asks individuals whether they were ill during a defined period and the number of days associated with disability days and with use of medical care. Aday et al. (1980) expanded on the approach by asking about the results of care and categorizing the type of illness as either requiring or not requiring care (according to professional judgment) and the degree to which the patient worried about the problem. The major disadvantage of the method is that it is cumbersome for respondents with complex illnesses. Although there is no validated method for deriving a summary index of appropriate response, Yergan et al. (1981) suggested a mechanism for categorizing the data to devise one. When outcome is included in the analyses, access is considered inadequate even if the outcome was good but care was not provided for problems that engendered a great deal of patient worry.

It should be noted that the last three methods described (use/disability, symptoms/response, episode of illness) make inferences about access from measures that involve utilization. In the terminology of Aday et al. (1980, p. 35), these measures reflect "realized access." It is important to recognize that these measures of access do not relate specifically to the source of primary care, although they could be adapted to do so.

Access to care may also be assessed at the *facilities (provider) level.* All of the methods used at the population level can also be applied to patients coming to the health care facility. In interpreting findings, however, these methods underestimate the extent of difficulty in access to services since patients with access problems that limit their use are underrepresented among the patients available for interview.

Other methods are particularly suited to use at the facilities (provider) level:

In the *simulated patient* approach individuals are trained to make an appointment by phone and then observed in their visit to the facility. Two types of measures are used, the urgency ratio and the frustration ratio. To determine the urgency ratio, a panel of individuals consisting of health professionals, administrators of the facility, and patients decides ahead of time whether particular complaints are critical, serious, chronic, or routine. These designations are based on the maximum time that should elapse before the patient is seen for each category of symptoms. Then, for each simulated patient with the problem, a ratio of the actual time to the appointment divided by the appropriate time to the appointment is calculated for patients in each category of condition.

The frustration index has three components: time spent on the telephone to make the appointment, time from arrival in facility till the patient is seen, and time spent waiting for laboratory tests. The appropriate time (in minutes) for each of these components was also set by the panel, and data from actual patients, regardless of their condition, is monitored and compared with the standards.

Facilities may set their own standards for performance on either of these indices. Alternatively, facilities may be compared with each other to determine which performs better.

In a second method of assessing accessibility (Weiner and Starfield, 1982), professionals are asked a series of questions about features of their office arrangements associated with accessibility, including the following:

- The availability of emergency appointment slots.
- The time lags between request and appointment for acute but nonemergency problems and for nonacute problems.
- The average duration of time in the waiting room.
- The availability of house calls.
- The use of an answering service when the office is closed.
- Provision for after-hours coverage.
- The availability of a sliding fee schedule related to ability to pay.
- The acceptance of payment from Medicare for eligible individuals.

The method assumes that all these features of access ought be provided, but its major use is comparative: to assess differences in accessibility to care across types of physicians and physician groups. When aggregated for an area, however, it could be used to compare the system's accessibility in different areas.

In a third approach to assessing accessibility, standards are set for each of several aspects of access, and then a facility or practice is evaluated (or evaluates itself) against the standards (Institute of Medicine, 1978). For accessibility, sixteen questions are posed. The first six relate to accessibility directly, the next three address convenience, and the remaining seven address acceptability:

> Is access to primary care services provided twenty-four hours a day, seven days a week?
> Is there an opportunity for a patient to schedule an appointment?
> Are scheduled office hours compatible with the work and way of life of most of the patients?
> Can most (90 percent) of medically urgent cases be seen within one hour?
> Can most patients (90 percent) with acute but not urgent problems be seen within one day?
> Can most (90 percent) appropriate requests for routine appointments, such as preventive exams, be met within one week?
> Is the practice unit conveniently located, so that most patients can reach it by public or private transportation?

Is the practice unit so designed that handicapped or elderly patients are not inconvenienced?

Does the practice unit accept patients who have a means of payment, regardless of source (Medicare, Medicaid)?

Is the waiting time for most (90 percent) of the scheduled patients less than one-half hour?

If a substantial minority (25 percent) of patients have a special language or other communication barrier, does the office staff include people who can deal with this problem?

Are waiting accommodations comfortable and uncrowded?

Does the practice staff consistently demonstrate an interest in and appreciation of the culture, background, socioeconomic status, work environment, and living status of patients?

Is simple, understandable information provided to patients about fees, billing procedures, scheduling of appointments, contacting the unit after hours, and grievance procedures?

Are patients encouraged to ask questions about their illness and their care, to discuss their health problems freely, and to review their records, if desired?

Does the practice unit accept patients without regard to race, religion, or ethnic identity?

Measuring Use of the Facility When Care Is First Believed to Be Needed

Both first contact and longitudinality involve consideration of the same process of care: utilization. The feature of utilization relevant to first contact care is the extent to which the *first* visit for a new problem is made to the regular source of care. In the case of longitudinal care, utilization should *always* occur (except for referral visits) at the source of primary care, regardless of the stage of or type of the particular problem.

In a health services system (such as that in the United States), where most people are not enrolled with a source of primary care, the measurement of first contact should always be made at the population level rather than at a facility level. The reason for this is that measurements at the facility level systematically exclude individuals who do not seek services because of poor access or for some other reason. Individuals with *few* visits to the facility are also underrepresented in data collected by taking a random sample of its users. Studies of access or of utilization conducted in facilities by interviewing patients exclude or underrepresent those who are most likely to have poor access or underutilization, unless the facility uses a roster of its patients to select samples for interviewing. For this reason population-based studies of first contact care are preferred to facility-based ones.

Measurement of the behavioral component of first contact care involves eliciting information concerning the place at which care is sought. This information might be obtained from such sources as medical records, claims forms, or billing data, but one cannot be sure that they encompass all uses of services. Another drawback to this technique is that these types of records usually do not indicate whether the visit was at the patient's initiative or resulted from a referral. Since first contact care pertains only to visits that patients initiate, the absence of this information fails to indicate accurately the extent to which first contact care is achieved.

One alternative is to ask individuals where they received care when they last sought it, what was the source of care for visits over a prior period of time, and whether the visit was made on the advice of another physician. Another alternative is to explore both accessibility and use of services at the same time, through questions such as the following:

> What is your regular source of care, that is, the place you usually go when you have a new problem you think requires a doctor's attention?
> Where did you go the last time you had a problem you thought needed a doctor's attention?
> How easy was it to obtain help from the doctor to whom you went? (Various aspects of access might be explored through branching of this question.)
> What was the reason you did not go to your regular source of care at this last visit? (The choice of reasons would include several clearly related to features that enhance accessibility to services and facilitate utilization.)

This series of questions could also be asked for visits before the last one, or even for all visits over a period of time such as two weeks, a month, or a year.

What Is Known About Attaining First Contact Care

Accessibility to Primary Care

There are several examples of how the measures previously described have been used to provide information on access to care and the adequacy of first contact in primary care.

The 1970s saw marked improvement in access to care as a result of the passage of the War on Poverty legislation in the mid-1960s. However, in the early 1980s various reductions were made in third-party coverage for care, especially for those unable to afford it. A national telephone survey conducted in the mid-1980s showed how access to services changed as a result of these reductions in insurance coverage. There was a marked increase in the number of people who reported they had no regular source of care and consequently had far fewer visits to physicians as well as hospitalizations (Freeman et al., 1987). The situation was especially marked among individuals who had no insurance; 31 percent lacked any regular source of care in 1986 (The Robert Wood Johnson Foundation, 1987). Although the reductions in use of care were not necessarily restricted to first contact care, the decrease in number of individuals who reported no regular source of care suggests that first contact care from a primary care source was seriously compromised as a result of reduced insurance coverage.

Different specialties differ in the degree to which their practices are accessible. Studies (Weiner, 1981; Starfield et al., 1973; Cherkin et al., 1986) indicate the following:

> Family physicians are less likely to require appointments for visits than are other physicians.

Family physicians are more likely to have regular office hours during weekends than are general internists.

Family physicians and internists are more likely to report that they make house calls or special visits to emergency rooms and nursing homes and that they make more such visits than other types of physicians.

Patients of family physicians wait less time for nonurgent appointments than patients of other types of physicians.

The offices of generalists and non-board-certified pediatricians or internists are more accessible than are those of board-certified pediatricians, internists, or other specialists.

National surveys indicate that people who report no regular source of care make substantially fewer visits than they need, as measured by the symptoms response ratio. People who have specialists as their regular source of care report making more visits that experts judged to be unnecessary. In contrast, those with a clinic or a generalist as their source of care have approximately the same number of visits overall as experts judged to be needed (Taylor et al., 1975).

Use of Primary Care for First Contact

The rate of visits to specialists in some organizational settings is much greater in some than in others largely because some plans allow patients to make appointments directly with specialists whereas others require a referral from a primary care physician (Perkoff, 1978; Starfield, 1986). Thus, in at least some facilities, first contact care is not achieved by primary care physicians. Whether the situation is a result of patient choice to go directly to the specialist where permitted to do so or a result of lack of accessibility or interest of the primary care physician cannot be determined from the data; only subsequent study will elucidate the reason for the failure of first contact in these primary care settings.

In a multisite study of use of services, people who reported a generalist as their source of care were much more likely to see that physician first when they had an acute illness than were people who reported a specialist as their regular source of care (Spiegel et al., 1983).

A national study of a representative sample of office-based physicians showed that general and family physicians had a higher proportion of first visits (in which the patient was not seen before) than pediatricians did. Pediatricians, in contrast, had a higher proportion of patients who had been seen previously but for other problems than the one occasioning the visit (Fishbane and Starfield, 1981). The same was the case for generalists compared with other types of physicians, although the differences between generalists and internists, ophthalmologists, and otolaryn-gologists were not large (Puskin, 1977). These findings suggests that generalists may serve more as the point of first contact for new problems than pediatricians, internists, or other specialists do.

Thus, the few studies that have examined the extent of first contact care indicate that individuals whose regular source of care is a primary care physician are more likely to be receiving first contact care from that physician than are individuals whose regular source of care is another type of physician.

Summary

First contact care involves the provision of services that are accessible (a structural feature of care) and the utilization of those services when a need for care arises (a process feature of care).

Information regarding the accessibility of care should be ascertained at both the population level and the facilities level.

The best way to elicit information concerning the attainment of first contact care is either by asking individuals in a population or by analyzing data on *all* visits made by the individuals in the population, to determine where individuals went when they first sought care for a new problem or need for care.

Accessibility of care varies with the type of physician. In general, primary care physicians, especially family physicians, are more accessible than are other types of physicians, particularly with regard to flexibility both in providing care without appointments and in the time of appointments.

Access to care is important in reducing mortality and morbidity. The use of primary care practitioners rather than specialists for first contact care is likely to lead to more appropriate care and better health outcomes.

The increasing focus on *gatekeepers* should be accompanied by a strategy to obtain information concerning the nature and extent of referrals, the development of better criteria for referral, and the appropriate roles of specialists in the care of patients.

References

Aday LL, Andersen R, Fleming G. Health Care in the U.S. Equitable for Whom? Beverly Hills, Sage Publications, 1980.

Andersen R. Health status indices and access to medical care. Am J Public Health, 1978; 68:458–63.

Catlin R, Bradbury R, Catlin R. Primary care gatekeepers in HMOs. J Fam Pr 1983; 17:673–78.

Cherkin D, Rosenblatt R, Hart LG, Schleiter M. A comparison of the patients and practices of recent graduates of family practice and general internal medicine residency programs. Medical Care 1986; 24:1136–50.

Christensen B, Sorensen H, Mabeck C. Differences in referral rates from general practice. Family Practice 1989; 6:19–22.

Dershewitz R (ed). Ambulatory Pediatric Care. Philadelphia, J.B. Lippincott Company, 1988.

Donabedian A. Aspects of Medical Care Administration. Specifying Requirements for Health Care. Cambridge, Mass., Harvard University Press, 1973, pp. 419–73.

Edwards WS and Berlin M. Questionnaires and Data Collection Methods for the Household Survey and the Survey of American Indians and Alaska Natives. Department of Health and Human Services, Public Health Service. National Center for Health Services Research and Health Care Technology Assessment. DHHS Publication No. (PHS) 89-3450, 1989.

Fihn S and Wicher J. Withdrawing routine outpatient medical services: effects on access and health. J Gen Intern Med 1988; 3:356–62.

Fishbane M and Starfield B. Child health care in the United States: A comparison of pediatricians and general practitioners. N Engl J Med, 1981; 305:552–56.

Freeman H, Blendon R, Aiken L, Sudman S, Mullinix C, and Carey C. Americans report on their access to health care. Health Affairs 1987; 6:6–18.

Gorell A, May L, and Mulley A. (eds). Primary Care Medicine. Philadelphia, J.B. Lippincott Company, 1981.

Hurley R, Freund D, and Taylor D. Emergency room use and primary care case management: evidence from four Medicaid demonstration programs. Am J Public Health 1989; 79(7):843–47.

Institute of Medicine. A Manpower Policy for Primary Health Care: Report of a Study. Washington, D.C., National Academy of Sciences, 1978.

Kovar MG. Data Systems of the National Center for Health Statistics. National Center for Health Statistics. Vital & Health Statistics 1(23), DHHS Publications No (PHS)89-1325, 1989.

Lawrence R and Dorsey J. The generalist–specialist relationship and the art of consultation. In: Noble J (ed). Primary Care and the Practice of Medicine. Boston, Little, Brown & Company, 1976.

Levinson DF. Toward full disclosure of referral restrictions and financial incentives by prepaid health plans. N Engl J Med 1987; 317(27):1729–31.

Lurie N, Ward N, Shapiro M, Gallego C, Vahaiwalla R, and Brook R. Termination of Medi-Cal benefits. A follow-up study one year later. N Engl J Med 1986; 314:1266–68.

Martin D, Diehr P, Price K, and Richardson W. Effect of a gatekeeper plan on health services use and charges: a randomized trial. Am J Public Health 1989; 79(12):1628–32.

Moore S. Cost containment through risk-sharing by primary-care physicians. N Engl J Med 1979; 300:1359–62.

Moore S, Martin D, and Richardson W. Does the primary-care gatekeeper control the costs of health care? Lessons from The SAFECO Experience. N Engl J Med 1983; 309:1400–04.

Penchansky R and Fox D. Frequency of referral and patient characteristics in group practice. Medical Care 1970; 8:368–85.

Penchansky R and Thomas JW. The concept of access: definitions and relationship to consumer satisfaction. Medical Care 1981; 19:127–40.

Perkoff G. An effect of organization of medical care upon health manpower distribution. Medical Care 1978; 16:628–40.

Puskin D. Patterns of Ambulatory Medical Care Practice in the United States: An Analysis of the National Ambulatory Medical Care Survey. Unpublished doctoral dissertation, The Johns Hopkins University, May 1977. p. 57.

Reagan M. Physicians as gatekeepers: a complex challenge. N Engl J Med 1987; 317:1731–34.

Richardson W. Measuring the urban poor's use of physicians' services in response to illness episodes. Med Care 1970; 8:132–42.

The Robert Wood Johnson Foundation. Access to Health Care in the United States: Results of a 1986 Survey. Special Report Number 2. Princeton, N.J. The Robert Wood Johnson Foundation, 1987.

Roland M and Morris R. Are referrals by general practitioners influenced by the availability of consultants? Brit Med J 1988; 297:599–600.

Roos N. Who should do the surgery? Tonsillectomy-adenoidectomy in one Canadian province. Inquiry 1979; 16:73–83.

Simon H, Reisman A, Javad S, and Sachs D. An index of accessibility for ambulatory health services. Medical Care 1979; 17:894–901.

Spiegel J, Rubenstein L, Scott B, and Brook R. Who is the primary physician? N Engl J Med 1983; 308:1208–12.

Starfield B. Effectiveness of Medical Care: Validating Clinical Wisdom. Baltimore, Md, The Johns Hopkins University Press, 1985.

Starfield B. Primary care in the United States. Int J Health Services 1986; 16:179–98.

Starfield B. Special responsibilities: the role of the pediatrician and the goals of pediatric education. Pediatrics 1983; 71:433–40.

Starfield B, Bice T, Schach E, Rabin D, and White KL. How "regular" is the "regular source of medical care?" Pediatrics 1973; 51:822–32.

Taylor D, Aday L, Andersen R. A social indicator of access to medical care. J Health Soc Behav 1975; 16:39–49.

Taylor RB. Categories of care in family medicine. Fam Med 1981; 13:7–9.

Weiner J. The Baltimore City Primary Care Study: An Analysis of Office-based Primary Care. Doctoral Thesis, The Johns Hopkins University School of Hygiene and Public Health, 1981.

Weiner J and Starfield B. Measurement of the primary care roles of office-based physicians. Am J Public Health 1982; 73:666–71.

Wilkin D and Dornan C. GP Referrals to Hospital: A review of research and its implications for policy and practice. Center for Primary Care Research, University of Manchester, July 1990.

Wilkin D and Smith A. Explaining variation in general practitioner referrals to hospital. Fam Pract 1987; 4:160–69.

Yergan J, LoGerfo J, Shortell S, Bergner M, Diehr P, and Richardson W. Health status as a measure of need for medical care: A critique. Med Care 1981; 19 (Suppl):57–68.

4

Longitudinality and Managed Care

I wish my own doctor could have been the one that took charge of everything. He knows me and we can talk together and understand each other.
Mrs. P, after the birth of her baby, who was normal but had an extensive workup for transient symptoms

This chapter describes longitudinality, its benefits, and its policy implications. It then turns to a discussion of how longitudinality may be measured and what is known about its achievement.

The achievement of primary care implies that one place, one individual, or one team of associated individuals serves as the source of care over a defined period of time, regardless of the presence or absence of particular health-related problems or of the type of problem. Having longitudinal care means that individuals in the population identify with a source of care as "theirs," that the provider or groups of providers at least implicitly recognize the existence of a formal or informal contract to be the "regular source of care," and that this relationship exists for a defined period of time or indefinitely until explicitly changed.

There is no useful distinction between identification of one physician, one team of physicians, or a particular place of care as the locus of longitudinality. However, when a team or a place is the source of longitudinality, the burden of coordination is likely to be greater than if a particular individual is the source, because patients are likely to be seen by more practitioners. Conversely, when an individual is the source of longitudinality, the challenges of providing first contact care and comprehensiveness are greater, since it is more difficult for an individual than for a team or medical organization to always be available and provide or arrange for a variety of needed services.

Continuity of care is a phrase that is often used to describe the extent to which patients see the same practitioner or visit the same facility from one visit to another, or over a period of time. Continuity may be a characteristic of specialty care as well as of primary care. Chapter 6 discusses more fully the differences between longitudinality and continuity.

The essence of longitudinality is a personal relationship over time, regardless of the type of health problems or even the presence of a health problem, between patients and a physician or a team of physicians and nonphysician personnel. Through this relationship the practitioners come to know patients over time, and patients come to know the practitioners. The benefits of this knowledge would be expected to accrue in a variety of ways. For example, patients should make fewer visits because many problems can be managed on the phone. Fewer hospitalizations should also result, since practitioners are more likely to be able to ascertain whether or not the problem could be managed at home.

> I've just been going from hospital to doctors' offices over and over again and I've been asking questions. The pains started after I had my operation. The doctors all take me from test to test—any test they can find, they give it to me, from a barium enema to an upper to GI series to whatever. And I just keep asking questions, and they just kept writing me prescription after prescription, telling me to take the medicines, which to this day haven't stopped the pain. I ask them "Why are you giving me all of these prescriptions when I'm so allergic to medicines?" When I asked them all these questions, they just tell me to trust them. So, at the beginning I took all the prescriptions which haven't helped. I got tired of those doctors who were "helping me" and who were telling me what I was feeling and not listening to how I said I was feeling. You just have to tell them to "hold it a minute." So I changed doctors because I wanted a second opinion. When I went to this other doctor, he examined me and then he just shook his head, and he said "I don't know, but take this medicine." He wrote me a prescription to take the medicine and told me to come back in two weeks. The pain wasn't getting any better and the medicine wasn't doing any good. So I went back to him, and he examined me again, and I asked him for the results of the tests he had taken, but he said he hadn't gotten them yet. He just sat there about five minutes, shaking his head, and the he writes me five prescriptions! And when I asked him what they were for, he said "Just take them. Trust me and take them, and we'll get to the bottom of this." I said "ok" and then he said that he was going to send in another gentleman to examine me, and that I should come back in about five weeks. So I sat there about five minutes until the other doctor came in. He took out his stethoscope and listened to one side—my right side—and he turned around to the desk and wrote me two more prescriptions. When I asked him what they were for, he said to just trust him and come back in two weeks. I told him that I wasn't a guinea pig and I wasn't going to take all of the prescriptions. I tore up all the prescriptions and left the office.
>
> Finally I went back to the family doctor who has been taking care of me for twenty years. She knows me really well and she was wonderful. She listened to me and didn't try to tell me what I was feeling. She took me off all the medications and the pains started going away.   Mrs. M, a fifty-year-old waitress

As with each of the four attributes of primary care, longitudinality is related to the others. The relationship with a regular source of care implies that this will be the place of first contact care. It also implies, as is noted in the next two chapters, that the regular source of care will ensure comprehensive, coordinated care.

The Benefits of Longitudinality

The benefits of association with a regular source of care have been documented in a wide variety of types of studies. Some people might claim that having a place (e.g., a clinic or a neighborhood health center) as the regular source of care is not equivalent to having a physician as a regular source of care. Studies of the benefits of longitudinality have been carried out using either the place or the physician as the basis for study, but there are few comparisons of the two. The following examples are divided according to whether the source was a place ("facility") or a physician. In most of the examples, actual association with a regular source of care (rather than the mere reporting of its presence) was present.

Studies Where the Source of Care Was a Facility

Children who receive their care from an identifiable regular source are more likely to receive preventive care than other children. Alpert et al. (1976) showed that children assigned randomly to a particular facility for all of their care had much higher rates of visits for health supervision and lower rates of visits for illness care than patients not enrolled in the facility, even though their prior rates for both types of visits were similar. McDaniel et al. (1975) showed that children receiving all of their care from a pediatric practice were more likely to have complete immunizations than were children receiving care from both a private pediatrician and a health department clinic. (However, the findings were not consistent across the four groups of pediatricians studied, indicating that the finding is not generalizable to all practices.)

Patients who make more visits to a facility during a period of time are more likely to receive indicated preventive care. This has been shown to be the case for children (Benson et al., 1984) and for adults, who benefited more from computerized reminder systems for preventive care if they had more rather than fewer visits (Chambers et al., 1989).

People who have a source of primary care are more likely to keep their follow-up appointments. Spivak et al. (1980) showed that children who had a source of primary care were twice as likely to keep follow-up appointments than were children without such a source. They were at least 50 percent more likely to do so than children whose primary care source was used for routine care and minor illnesses but who were taken to the emergency room for more serious problems, or children who used more than one primary care source, with or without emergency room use.

People whose care is longitudinal have fewer emergency hospitalizations and shorter hospitalizations, as demonstrated by a study in which men over the age of fifty-five were randomly assigned to two different groups for care, one longitudinal and the other not (Wasson et al., 1984). Nationally, children have fewer emergency room visits if they have a regular source of care than if they do not (Orr et al., 1991).

Inner city children are more likely to receive care for undifferentiated episodes of illness, for earaches, and for regular care for asthma if they had a longitudinal source of care then if they did not have such a source (German et al., 1976; Salkever et al., 1976).

People whose care is provided longitudinally are less likely to contract preventable illnesses, as was demonstrated in a study of the occurrence of acute rheumatic fever in children in Baltimore (Gordis, 1973). Longitudinality of care also may reduce the likelihood of poor outcomes, especially low birth weight. When prenatal care is a continuation of ongoing care, infants are more likely to be of normal birthweight than if their prenatal care was not part of a program of care that antedated pregnancy (Starfield, 1985).

Costs of care are reduced for individuals receiving longitudinal care. An analysis of national data by Butler et al. (1985) showed that children who reported a regular source of care had approximately 25 percent lower total costs of care than children with no regular source of care; costs for the latter group of children had a much larger standard error (indicating greater heterogeneity of experiences) than those for the children with a regular source of care.

The relationship between longitudinality of care and use of services depends on how the issue is investigated. Those studies in which the regular source of care is identified retrospectively—that is, after the utilization has occurred—indicate that the presence of a regular source of care is associated with an increased number of visits (Andersen and Aday, 1978; Scitovsky et al., 1979; Marcus and Stone, 1984). Studies in which the regular source of care is identified before utilization indicate that having a regular source of care reduces utilization (Breslau and Reeb, 1975; Alpert et al., 1976). The reason for this is that relatively recent use of services is likely to generate a perception of a regular relationship; individuals with greater use are more likely to report that the physician who provided the care is the regular source of care. Prospective studies, therefore, provide a more accurate assessment of the impact of longitudinality on use of services; they show that frequency of services, especially for illness care, is reduced.

Studies in Which the Regular Source of Care Was an Individual Practitioner

Longitudinality facilitates compliance with prescribed medication. In a study in upstate New York, more medication was taken among children whose own physician prescribed it; the longer the duration of relationship with the physician, the greater the compliance (Charney et al., 1967). The same was the case for children being treated for otitis media (Becker et al., 1972).

Patients who had a regular source of care but did not see that physician were more likely to be judged as not needing care by the physician who saw them (not necessarily their regular source) for a problem than if their own physician had made the judgment. This finding suggests that physicians who have developed a relationship with patients are better able to appreciate the patients' needs than physicians who are not familiar with them (Steinwachs and Yaffe, 1978).

Longitudinality of care facilitates the recognition of certain types of problems. For example, Becker et al. (1974) demonstrated that behavior problems in children randomly assigned to receive care by a team consisting of a physician, nurse, and registrar, which remained constant over time, were more likely to be recognized than those in children seen in the same facility by the next provider available.

Patients seen for consultation in a medical center are more likely to return to the referring physician if they consider that physician to be their regular source of care (Lawrence and Dorsey, 1976). That is, identification of a source of longitudinality improves the extent to which patients complete the referral process by returning to the referring physician.

When patients visit the same practitioner rather than different practitioners, care is more efficient. A population-based study involving review of all claims of a random sample of Medicaid patients (ages zero to twenty-one) for three years revealed that longitudinality was associated with a reduction in hospital admissions and overall costs (Flint, 1987).

Longitudinality is associated with increased satisfaction on the part of patients, as shown in several studies (Wasson et al., 1984; Becker et al., 1972, 1974).

What accounts for the beneficial effects of longitudinality? Although the specific mechanisms have not been explored, there are several possibilities. Through time, a sense of trust in the physician may make patients more comfortable in divulging relevant information and more responsive to their physician's recommendations. Similarly, physicians involved in longitudinal care may be more sensitive to relatively subtle cues that help elucidate the nature of the patient's problem. Accumulated knowledge about the patient's background and illness experiences may aid the physician in arriving at a more accurate assessment of the patient's problem. The physician as a "repository of information" is the basis for the notion that continuity of care is a useful aspect of health services; this subject is addressed in Chapter 6.

Concern has occasionally been expressed that longitudinality may delay the recognition of serious problems (Evang, 1960; Miller, 1973), but little data bear on this subject.

Policy and Research Implications of Longitudinality: Managed Care

In current parlance, the essence of longitudinality is encompassed by the term *case management,* a concept long part of social work which is now being applied in the health services arena. In its original sense (and as used in the Allied Health Services Act of 1972), case management was a technique to improve access to services by encouraging community support services to coordinate benefits and maximize their attainment. With the Omnibus Reconciliation Act of 1972, however, case management became a tool to control access to certain services by "locking" patients into one source of care, which would be responsible for all other use (Spitz and Abramson, 1987). Inherent in both conceptualizations, however, is consideration of all aspects of the patient's care for a period of time. Its essence is knowledge of the patient in the context of the variety of needs that arise over time. Thus, it is most closely related to that element of primary care that is patient and time focused, that is, longitudinality.

The concept of case management in the context of health services starts with the premise that case management is composed of a set of functions that can be ordered along a spectrum in two dimensions: patient-oriented versus self-oriented and clinical versus management skills (Hurley, 1986). Functions that are primarily patient-oriented and clinical include advocate, agent, confidant, caregiver, healer, and interpreter. Functions that are patient-oriented but managerial include broker, guide/escort, educator, service evaluator, and investment counselor. Functions that are primarily clinical and directed at self-interest include clinician, diagnostician, therapist, risk manager, and researcher/scientist. Those that are directed primarily at self-interest but are managerial in type include auditor/controller, gatekeeper, prior authorization, purchasing agent, and resource allocator. In this conceptualization, several other features of primary care are subsumed under a case management function, notably coordination (a patient-focused function that bridges the clinical and management functions) and first contact care (a self-interest function that is managerial).

Because case management encompasses so many functions, someone other that a physician may be required to fulfill the role. Liptak and Revell (1989) provided empirical evidence to support this suggestion. The researchers first defined case management as identifying and assessing the needs of children and their families; planning and arranging for medical and nonmedical services; facilitating and coordinating services (including the training of community providers); monitoring services and patient progress (follow-up); and counseling, educating, training, and supporting patients and their families (empowerment). The researchers then surveyed family physicians and pediatricians working in medical center specialty clinics and parents of patients with chronic illness from the same clinics. About 60 percent of the physicians thought the primary care physician should be the case manager, 20 percent thought parents were more appropriate for the role, 15 percent suggested a specialty clinic or physician, and 5 percent suggested a community health nurse. Physicians were unsure who was providing case management for 51 percent of their own chronically ill patients. Specific considerations that physicians thought would be most important in their own role as case manager were diagnosis, severity of the illness, and resources available in the specialty area. Parents, however, rated specialty clinics and primary care physicians almost equally knowledgeable about their child's health, as well as equally accessible. Both parents and physicians rated specialty clinics or specialty physicians as better informed than primary care physicians about community services. Physicians differed from parents in their rating of priorities for information; they underestimated parental need for information about the child's diagnosis, treatments being prescribed, and prognosis but overestimated their need for information regarding financial aid, insurance, and vocations.

Thus, it seems likely that the most needed aspects of case management, at least from the patient's viewpoint, are closer to the clinical than to the managerial aspects of case management. Considering the nature of training of physicians, it makes more sense to emphasize case management functions that are clinical, reserving the managerial function for others working in concert with physicians. That is, those functions that require clinical knowledge should be reserved for the physician whereas those that can be performed by others in concert with clinicians should be delegated to them. The clinical functions are those that concern recognition of the patient's health needs, diagnosis, therapy, and reassessment of needs, that is, the functions of primary care that require the clinical judgment that derives from a longstanding knowledge of the patient and the patient's needs.

The approach taken by the New York State Department of Social Services and Health (Child Health Financing Report, 1990) is consistent with this thrust. In this program for paying physicians who care for low-income children, physicians must provide for twenty-four-hour coverage of the practice, arrange for hospital admissions, specialty consultations, and ancillary services, and adhere to guidelines for preventive care. Specialists must also provide summaries of consultations to the primary care physician and inform the latter when they arrange hospital admissions. Thus, managed care consists of functions pursuant to achieving all of the essential features of primary care through the mechanism of longitudinality.

Financing for case management was incorporated into public policy with the passage of section 9508 of the Consolidated Omnibus Budget Reconciliation Act of 1985, which provided funds for the provision of case management within the Social Security Act (Medicaid). Similarly, Public Law 99–457 (the Early Intervention Program for Infants and Toddlers with Handicaps) mandates case management services for children who are part of the program. Further empirical study is required before the reasonable bounds of case management are defined. The imperatives of new forms of organization of health services may increasingly impose nonclinical managerial roles on primary care physicians (Hillman, 1987). There is, at present, no basis for judging the ability of physicians to enlarge their scope of skills to encompass the functions that are primarily managerial and nonclinical, and to judge whether this expansion of scope produces desired outcomes in terms of patients' health as well as the costs of care.

Measurement of Longitudinality

Central to the measurement of longitudinality is the notion that individuals should be able to identify their source of care and facilities should be able to identify their eligible population (the "structural" feature). Furthermore, individuals should use the regular source for all problems except those for which the primary care physician refers them elsewhere (the "process" feature). Although there is a relationship between the use of a regular source of care for all problems and for first contact, the concepts are different. For example, an individual might have sought care from the "primary care" physician for new episodes of chest pain and earache at separate times during a year, thus satisfying the criteria for first contact care, but independently (without a referral from the primary care physician) also seek ongoing care from another physician for chronic sinusitis (thus not achieving longitudinality).

This section first discusses ways of measuring the regular source of care and the eligible population, and then ways of measuring appropriate use of the regular source of care.

Identification of a Regular Source of Care and Defining an Eligible Population

At the *population level* the extent to which individuals have a regular source of care is often ascertained by means of a household survey, such as the National Health Interview Survey. This ongoing survey of a nationally representative sample of households periodically assesses whether people have a regular source of care and whether that source is a particular physician or a place. Data are available to determine time trends in reporting on a regular source of care as well as changes in the type of source.

Over the years, researchers and pollsters have conducted many ad hoc surveys, some of them periodic. Among the most extensive are those conducted at the University of Chicago (Andersen et al., 1976; Aday et al., 1980, 1984). More recently, the Robert Wood Johnson Foundation (through the Harris polls) has conducted three waves of interview by telephone: in 1976, 1982, and 1986 (Freeman et al., 1987).

Several surveys conducted during the 1970s consistently showed that about 85 to 90 percent of the population indicated they had a regular source of care (National Center for Health Statistics, 1985). The 2 percent who indicated that they had more than one doctor, depending on the medical problem, and the 13 percent who said they did not have a physician or place that is a regular source of care lacked at least the aspect of primary care that concerns longitudinality. The extent of identification with a regular source of care decreased during the early 1980s, so that by the mid-1980s, 18 percent of the population reported no regular source of care (Freeman et al., 1987). During the same time period, use of ambulatory services declined. As will be shown later, identification of a regular source of care is heavily influenced by whether or not services have been used.

Compared with the whole population (of which 18 percent have no regular source of care), the following subgroups are less likely to have a regular source of care: males (22 percent), those in urban areas (19 percent), blacks and Hispanics (20 percent and 30 percent, respectively), poor families (20 percent), the uninsured (31 percent), and those with chronic disease (20 percent) (Freeman et al., 1987).

At the *facilities level,* identification with a regular source of care is reflected in the existence of patient rosters which indicate that the practitioner or group identifies an eligible population for which it is responsible. In the United States, the existence of such rosters is uncommon although new forms of organizations of services that incorporate the "gatekeeper" concept (see the chapter on first contact care) will increase their frequency.

In many other countries, patient rosters are an integral part of primary care and often serve as the basis for payment of practitioners. To study the validity of the roster of a health center in the province of Ontario Canada, a mail and a telephone survey ascertained the extent to which patients and the facilities agreed that the facility was their regular source of care. The roster included all individuals who indicated they intended to use the practice as their regular source of care. For patients already known to the practice at the time the register was developed, those whose records indicated a utilization pattern compatible with regular use were included. The roster was augmented with names of family members identified by new patients. The facility was considered the regular source of care if survey respondents indicated that it was their "usual" source (Anderson et al., 1985). (The accuracy of the roster compared with the mail survey, in this one health center, was over 90 percent.)

The Appropriate Use of the Regular Source of Care

Research on *use* of the regular source of care has confused the concepts of longitudinal care, majority of care, and sequential care (Dietrich and Marton, 1982; Wall,

1981; Starfield, 1980). As a result, some of the methods described in this section have been referred to as "continuity" measures, whereas they really address the phenomenon of longitudinal care.

No standards for longitudinality have been proposed. Therefore, all of the measures described in this chapter are relative measures in that the results of one assessment are compared with another; higher score reflects better longitudinality.

Four measures are available for use at the *population level:*

1. The UPC (usual provider continuity). In this measure, the number of visits to the regular source of care is divided by the total number of visits in the same time period. The resulting ratio is known as the UPC (Breslau and Reeb, 1975). The closer the ratio is to 1, the higher the longitudinality is.

2. COC (continuity of care). The score on the UPC is very sensitive to the total number of visits; for example, if an individual makes only one visit in a year and it is to the regular source of care, the UPC is 1. The fewer the number of visits, the easier it is to achieve a high UPC. The COC corrects for this statistical problem by requiring detailed visit data and using it to correct for the effect of number of visits in its calculation. It also corrects for the total number of different providers that are seen, and considers visits to specialists that are made *on a referral* from primary care as a visit to the primary care physician. (Details are provided in Bice and Boxerman, 1977.)

3. LICON (likelihood of continuity). This measure corrects for the effect of number of sources available as well as the total number of visits. The greater the number of sources available, the greater the likelihood that practitioners other than the regular source of care will be seen. (Details are provided in Steinwachs, 1979.)

Each of the three measures described, when applied to all visits within a defined period of time (usually a year), indicates the extent to which a regular source of care is used over time. The measures do not determine the extent to which visits are for particular problems or for a full range of services. They also do not determine whether or not any of the visits were made upon referral. Therefore, in applying any of these measures to ascertaining longitudinality, it is important to eliminate visits resulting from a referral by the primary care provider to another provider.

4. Another measure of longitudinality takes into *consideration the nature of the problem for which care was sought.* Individuals are contacted by survey to determine their regular source of care. Claims, medical records, or patient interview are used to determine the nature of the care provided (optimally the reason for care, but usually the diagnosis given) and the place where it was provided. The extent to which nonreferred care was sought from the regular source of care (rather than from another source) for a variety of reasons is the measure of longitudinality. Although this measure has not yet been applied to its full potential, it was used in modified form in the Rand Health Insurance Experiment (Spiegel et al., 1983), a national demonstration of the effect of copayments on utilization and health status. In this application, individuals were asked to indicate the name of their "personal" physician, that is, the individual to whom test results from a scheduled multiphasic screening examination should be sent. Records were reviewed to determine the physician providing the plurality of visits as well as to determine the type of physicians patients consulted for upper respiratory infections, hypertension, and

general examinations. Only 12 percent of patients named specialists as their personal physician, and only 6 percent named physicians who were listed as specialists in the American Medical Association Masterfile and the American Medical Directory. When the primary physician was defined as the one involved in the largest number of visits, one in three patients was categorized as having a specialist as the primary care physician. When the primary physician was defined as the place where care was sought for common problems, approximately one in ten patients had a specialist as their source of primary care.

An appropriate modification of this method would be useful in determining the extent of longitudinality of care. In this method, patients are asked about their regular source of care, which can be defined as either the physician to whom they go for care when they have a new problem or the physician to whom they would want results of medical tests sent. Records of all visits over a period of time are reviewed to determine where the individual went for care for preventive or administrative reasons, for symptoms or signs of potential new illness, for care of existing problems, and for diagnostic or therapeutic reasons. Only visits that occurred on the patient's initiative are considered; that is, visits resulting from a referral by the regular source of care are excluded from consideration. When medical records or claims forms do not state whether the visit was patient-initiated, patients may be queried to determine where they went the last time they required a visit for these reasons and whether they went on referral from their regular source of care or on their own initiative. The percentage of unreferred visits for the variety of purposes that were to the regular source of care indicates the extent of longitudinality of care that has been achieved.

Information about the attainment of longitudinality at the population level requires obtaining information about the regular source of care by a population-based method, that is, by a method that contacts individuals independently of their visits to the facility. However, any of the preceding four methods could be calculated with information provided by patients when they appear at the facility. In this case, the measure would be considered facilities based rather than population based.

At the *facilities level,* methods of assessing longitudinality use information obtained from records within the facility. Therefore, the extent of longitudinality can be inferred only for those individuals who actually used the facility during the time period of interest.

As noted earlier, the UPC, COC, LICON, and place of visit, for a variety of reasons, are facility-based measures of longitudinality when the source of information is obtained from records or patients in the facility. The k index (Ejlertsson and Berg, 1984) is a measure of longitudinality at the facilities level. This measure is similar to the UPC (proportion of total visits to the regular source of care) except that the regular source of care is considered the individual seen by the patient at a particular visit rather than a physician identified by the patient as the regular source of care. The f index (Smedby et al., 1986) is similar to the k index in that it is the fraction of visits over a defined time that were made to a physician seen at a specified visit. A comparison of UPC, COC, and the k index with examples of different numbers and patterns of use of care is provided in Ejlertsson and Berg (1984).

What Is Known About the Attainment of Longitudinality

No known studies ascertain the nature of the regular source of care on a population level and compare it with information on appropriateness of use of services from the regular source of care at the same level. Following are population-based studies that provide information on the type of physician who serves as the regular source of care and facilities-based studies that determine the types of physicians who achieve longitudinality.

Various surveys indicate that generalists (family physicians, general practitioners, pediatricians, or general internists) are indicated as the regular source of care for the majority of the population. In the United States, where general internists and pediatricians as well as family physicians are considered to be primary care practitioners, family physicians and general practitioners account for about half of all physicians identified as the regular source of care; internists and pediatricians each account for about 14 percent. The remaining 20 percent of physicians identified as the regular source of care are specialists. For the population of children, pediatricians are identified as the regular source of care for 56 percent of one- to five-year-olds and 33 percent of six- to seventeen-year-olds. Income of the family is related to the type of physician indicated: lower-income families are more likely to identify a family physician or general practitioner and less likely to identify an internist or pediatrician than high-income families (Aday et al., 1980, p. 53). Moreover, even when low-income families identify a primary care specialist (such as an internist or pediatrician), the physician is less likely to be board certified (Starfield et al., 1973).

In the Rand Health Insurance Experiment (which excluded people over age sixty-two and the 7 percent of the population in the highest income groups), 88 percent of the population in three areas of the country named a primary care physician as their personal physician. Almost half (49 percent) named general practitioners or family physicians, 16 percent named internists, and 22 percent named pediatricians. About one in eight people (12 percent) named a specialist: 5 percent obstetricians/gynecologists, 4 percent surgeons, 2 percent internists–sub-specialists, and the remainder other types of physicians or physicians with unknown specialty (Spiegel et al., 1983).

Weiner (1981), in a study of the distribution and achievement of primary care within an entire metropolitan area (Baltimore), asked patients visiting a random sample of all types of physicians to indicate the year they first visited the physician. The duration of the physician patient relationship was calculated from this information and from the current year. The assumption was made that a longer duration would more likely reflect an ongoing relationship not associated with care of a particular problem or type of problem. Patients of both internists and family physicians reported relationships of approximately seven to eight years, on average, about double that reported by patients of other types of physicians (about three to four years on average); this suggests that generalist physicians provide more longitudinality than specialists.

The National Ambulatory Medical Care Survey, a periodic survey of a representative sample of office-based physicians in the United States, provides a means of assessing longitudinality in its determination of whether visits are first visits, follow-up visits (there was a previous visit for the same problem), or first visits for a particular problem by a patient previously seen for other problems. Puskin (1977) analyzed visits made by adults and calculated a longitudinality ratio for each type of physician. This ratio consisted of the number of patients seen before for other problems divided by the number of patients never seen before. General practitioners and family physicians had the highest ratios (2.36); general internists followed with 1.65. Several types of specialists had ratios slightly lower than that of internists (cardiovascular specialists at 1.54, general surgeons at 1.40, and obstetrician/gynecologists at 1.14) but the ratio of other specialists was much lower (.23 to .42).

A similar analysis was done for visits by children except that only pediatricians and general practitioners/family physicians were compared (Fishbane and Starfield, 1981). The longitudinality ratio for pediatricians (4.65) was much higher than that for general physicians (2.27), although the ratio of 2.27 for generalists was similar to that found for adult patients and much higher than the ratio for other types of physicians in the adult study.

The proportion of patients reporting one physician as responsible for their prenatal care and the number of different doctors seen during the course of pregnancy were compared for those receiving prenatal care from either family practice physicians or obstetricians in two different facilities in the same institution (Shear et al., 1983). In the family practice facility, 91 percent of patients reported one physician as responsible for care; the comparable figure for the obstetrician-staffed facility was 11 percent. The average number of physicians seen in the two facilities was 1.7 and 4.3, respectively, thus indicating that generalists are more likely to serve as a source of longitudinal care (at least during pregnancy) than specialists.

These findings indicate that generalists, pediatricians, and internists are more likely to provide longitudinal care than specialist physicians.

Studies that examine the extent to which physicians of different types provide care for a variety of problems also indicate that generalist physicians are more likely than specialists to provide longitudinal care. For example, in the study of Baltimore physicians, which asked patients where they last received a regular check-up and care for a cold or the flu, pediatricians, internists, and family physicians were named two to three times as often as the source of care for both types of problems as were specialist physicians (Weiner, 1981).

Investigators in the Rand Health Insurance Study (Speigel et al., 1983) came the closest to assessing the achievement of longitudinality by determining the percentage of people who saw the same physician for all visits, the majority of visits, and less than a majority of visits over a year, according to the type of physician who was the "majority" source of care (defined as the physician seen for the most visits, as determined from claims forms). Patients with a generalist as the majority source were more than twice as likely to have seen the same physician for all visits as were patients with a specialist as the majority source (36 versus 16 percent). Other analyses indicated that patients for whom a specialist was the majority source were much *less* likely to see that specialist for an upper respiratory infection (33 percent) than were patients for whom a generalist was the majority source; the latter saw that generalist 99 percent of the time. Even patients seen for a condition such as hypertension were more likely to be seen by their majority source if that source was a generalist than if it was a specialist. All patients whose source was a generalist saw that generalist for hypertension; only 53 percent of patients of specialists saw that physician for hypertension and 47 percent saw a generalist. These findings indicate that longitudinality is much more frequently a feature of the care of generalists than of specialists.

Summary

Longitudinality implies the existence of a regular physician or group of physicians and use of that source for care that is not limited to certain problems or types of problems. Its assessment involves measuring the structural feature by which people identify a regular source of care and physicians or facilities identify their eligible population, and the process of care that concerns appropriate use of that regular source of care.

There are several methods of assessing longitudinality of care, some population based and some facilities based.

Longitudinality of care is less well achieved by certain segments of the population, especially individuals in the lower social classes and other relatively disenfranchised groups.

Longitudinality is associated with a variety of benefits, including less use of services, better preventive care, more timely and more appropriate care, less preventable illness, greater satisfaction with care, and lower total costs.

Case management is a relatively new concept in health services. As a function that is focused on the patient over time, it is most closely related to the longitudinality feature of primary care. Its functions are both clinical and managerial, involving priorities that are focused on patients' needs as well as on professional imperatives. The functions that require clinical skills (such as first contact care, comprehensiveness, and coordination of care) are logical components of case management, but further research and evaluation is required to determine the feasibility and impact of assumption of more managerial functions by physicians.

References

Aday L, Andersen R, and Fleming G. Health Care in the U.S. Equitable for Whom? Beverly Hills, Sage Publications, 1980.

Aday L, Fleming G, and Andersen R. Access to Medical Care in the U.S.: Who Has It, Who Doesn't. Chicago, Pluribus Press, 1984.

Alpert J, Robertson L, Kosa J, Heagarty M, and Haggerty R. Delivery of health care for children: report of an experiment. Pediatrics 1976; 57:917–30.

Andersen R and Aday LA. Access to medical care in the U.S.: realized and potential. Medical Care, 1978; 16:533–46.

Andersen R, Lion J, and Anderson O. Two Decades of Health Services: Social Survey Trends in Use and Expenditure. Cambridge, Mass., Ballinger Publishing Co., 1976.

Anderson J, Gancher W, and Bell P. Validation of the patient roster in a primary care practice. Health Services Research 1985; 20:301–14.

Becker M, Drachman R, and Kirscht J. Continuity of pediatrician: new support for an old shibboleth. J Pediatr 1974; 84:599–605.

Becker M, Drachman R, and Kirscht J. Predicting mothers' compliance with pediatric medical regimens. J Pediatr 1972; 81:843–54.

Benson P, Gabriel A, Katz H, Steinwachs D, Hankin J, and Starfield B. Preventive care and overall use of services: Are they related? Am J Dis Child 1984; 138:74–78.

Bice T and Boxerman S. A quantitative measure of continuity of care. Medical Care 1977; 15:347–49.

Breslau N and Reeb K. Continuity of care in a university-based practice. J Med Educ 1975; 50:965–69.

Butler J, Winter W, Singer J, and Wenger M. Medical care use and expenditure among children and youth in the United States: analysis of a national probability sample. Pediatrics, 1985; 76:495–507.

Chambers C, Balaban D, Carlson B, Ungemack J, and Grasberger D. Microcomputer-generated reminders: improving the compliance of primary care physicians with mammography screening guidelines. J Fam Practice 1989; 29:273–80.

Charney E, Bynum R, Eldredge D, Frank D, MacWhinney J, McNabb N, Scheiner A, Sumpter E, and Ikor H. How well do patients take oral penicillin? A collaborative study in private practice. Pediatrics 1967; 40:188–95.

Child Health Financing Report. New York hikes Medicaid fees. Amer Acad Pediatr 1990; 7:1.

Dietrich A and Marton K. Does continuous care from a physician make a difference? J Fam Practice 1982; 15(5):929–37.

Ejlertsson G and Berg S. Continuity-of-care measures: an analytic and empirical comparison. Medical Care 1984; 22:231–39.

Evang K. Health Service, Society, and Medicine. London, Oxford University Press, 1960, pp. 87–88.

Fishbane M and Starfield B. Child health care in the United States. N Engl J Med 1981; 305:552–56.

Flint S. The impact of continuity of care on the utilization and cost of pediatric care in a Medicaid population. Unpublished doctoral dissertation, University of Chicago, 1987.

Freeman H, Blendon R, Aiken L, Sudman S, Mullinix C, and Corey C. Americans report on their access to care. Health Affairs 1987; 6:6–18.

German P, Skinner A, Shapiro S, and Salkever D. Preventive and episodic health care of inner-city children. J Comm Health 1976; 2:92–106.

Gordis L. Effectiveness of comprehensive care programs in preventing rheumatic fever. N Engl Med 1973; 289:331–35.

Hillman A. Financial incentives for physicians in HMOs: Is There a Conflict of Interest? N Engl J Med 1987; 317(27):1743–48.

Hurley R. Toward a behavioral model of the physician as case manager. Soc Sci Med 1986; 23(1):75–82.

Lawrence R and Dorsey J. The generalist–specialist relationship and the art of consultation. In: Noble J (ed). Primary Care and the Practice of Medicine. Boston, Little Brown & Company, 1976.

Liptak G and Revell G. Community physician's role in case management of children with chronic illnesses. Pediatrics 1989; 84(3):465–71.

Marcus A and Stone J. Mode of payment and identification with a regular doctor: A prospective look at reported use of services. Medical Care 1984; 22:647–60.

McCormick J. Evaluating primary care. In: Walter W Holland (ed). Evaluation of Health Care. Oxford University Press, Oxford, 1983.

McDaniel D, Patton E, and Mather J. Immunization activities of private practice physicians: a record audit. Pediatrics 1975; 56:504–07.

Miller M. Who receives optimal medical care? J Health Social Behavior, 1973; 14:176–82.

National Center for Health Statistics. B Bloom and SS Jack. Persons With and Without a Regular Source of Medical Care. United States, Vital and Health Statistics. Series 10 No. 151. DHHS Pub No. (PHS) 85-1579. Public Health Service. Washington, U.S. Government Printing Office, 1985.

Orr S, Charney E, Straus J, and Bloom B. Emergency Room Use by Low Income Children with a Regular Source of Health Care. Med Care 1991; 29:283–86.

Puskin D. Patterns of ambulatory medical care in the United States: An analysis of the National Ambulatory Medical Care Survey. Unpublished Doctoral Dissertation, The Johns Hopkins University School of Hygiene and Public Health, Baltimore, 1977.

Salkever D, German P, Shapiro S, Horky R, and Skinner A. Episodes of illness and access to care in the inner city: a comparison of HMO and Non-HMO populations. Health Services Research 1976; 1:252–70.

Scitovsky A, Benham L, and McCall N. Use of physician services under two prepaid plans. Medical Care 1979; 17:441–60.

Shear C, Gipe B, Mattheis J, and Levy M. Provider continuity and quality of medical care: A retrospective analysis of prenatal and perinatal outcome. Medical Care 1983; 21:1204–10.

Smedby O, Eklund G, Eriksson E, and Smedby B. Measures of continuity of care. A register-based correlation study. Medical Care 1986; 24:511–18.

Spiegel J, Rubenstein L, Scott B, and Brook R. Who is the primary physician? N Engl J Med 1983; 308:1208–12.

Spitz B and Abramson J. Competition, capitation, and case management. Milbank Mem Fund Q 1987; 65(3):348–70.

Spivak H, Levy J, Bonanno R, and Cracknell M. Patient and provider factors associated with selected measures of quality of care. Pediatrics 1980; 65:307–13.

Starfield B, Bice T, Schach E, Rabin D, White KL. How "regular" is the "regular source of medical care?" Pediatrics 1973; 51:822–32.

Starfield B. Continuous confusion? Am J Public Health 1980; 70:117–19.

Steinwachs D and Yaffe R. Assessing the timeliness of ambulatory medical care. Am J Public Health 1978; 68:547–56.

Steinwachs D. Measuring provider continuity in ambulatory care: An assessment of alternative approaches. Medical Care 1979; 17:551–65.

Wall EM. Continuity of care and family medicine: definition, determinants, and relationship to outcome. J Fam Practice 1981; 13(5):655–64.

Wasson J, Sauvigne A, Mogielnicki R, Frey W, Sox C, Gaudette C, and Rockwell A. Continuity of outpatient medical care in elderly men. A randomized trial. JAMA 1984; 252:2413–17.

Weiner J. The Baltimore City Primary Care Study: An analysis of office-based primary care. Unpublished thesis, The Johns Hopkins University School of Hygiene and Public Health, Baltimore, 1981.

5

Comprehensiveness and Benefit Packages

Alisa is a four-year-old child whose growth and development have been normal. For the past year, however, she has had acute otitis media which never seems to clear despite apparently adequate antibiotic therapy. Several doctor's appointments have not been kept and her mother does not call to cancel or reschedule them. At a monthly case conference, the situation was discussed and a decision was made to have Dr. S make a home visit.

The family lives in an apartment house in a low-income neighborhood. The building has an elevator but the corridors are dark. The family's apartment is, however, light and well kept. When Dr. S rang the bell, she was greeted warmly by Mrs. M and her three children, who were not in school that day. In the course of discussing the family and their health problems, Dr. S asked where the children slept, and Mrs. M volunteered to show her the apartment. As they sat talking in the living room over coffee, which Mrs. M had prepared, Dr. S noticed a closed door leading off the room and inquired about what it was. Mrs. M offered to show her and opened the door. Sitting in the corner was an elderly appearing white-haired woman, rocking back and forth and talking to herself. A few minutes of observation convinced Dr. S that the woman, the maternal grandmother, was actively hallucinating.

Subsequent social work intervention resulted in appropriate placement of the grandmother. Alisa's ear problem, which persisted because of Mrs. M's distraction with her own mother's schizophrenia and her consequent inability to focus on the prescribed therapy, resolved.

Comprehensiveness implies a broad range of services and recognition of the need to apply them directly or arrange for their provision when needed. These are the *structural* and *process* components, respectively. But what constitutes a "broad range of services"? What is optimal, or even acceptable, changes over time, expanding as new knowledge and improved technology widen the range of possibilities.

A comprehensive approach to primary care must, *at least,* involve the four steps of the medical process: problem (or needs) recognition, diagnosis, management, and reassessment. All medical care, even specialty care, involves all four steps, but primary care has an especially broad responsibility in the area of needs recognition, involving prevention as well as caring and healing. The major challenge for prevention is to recognize situations in which a preventive intervention is both needed and justified. The range of types of health problems in primary care is greater than in any other type of care, and therefore the range of preventive options is wider. Recognizing when preventive services are required is a challenge not only for those who are ill and appear for care, but also for those who are well and need encouragement to seek care. Since primary care is ongoing, the caring and healing functions are likely to be more prolonged, extensive, and far-ranging. More extensive types of resources must often be brought to bear. Services not normally provided in specialty care, such as home visits, other efforts at outreach, and marshalling a variety of health-related social services, are often required as part of comprehensive care. Since primary care deals with a broader array of health concerns and does so within a broader social context, it must have a wider array of types of resources at its disposal.

It seems self-evident that the greater the range of available services, the better the care will be. Such an assumption is not necessarily true, however; some types of services may not be effective, some may be effective but not worth their cost, and some may be harmful. As is noted later in this chapter, deciding on an appropriate range of services is an important policy issue.

In primary care the relevant policy issue is the range of services provided directly by the primary care service or indirectly by referral from primary care. In many (if not most) health systems, hospitalizations and the most expensive specialty services are covered by insurance plans or national health systems. In contrast, many primary care services are often not included, sometimes on the grounds that they are more "discretionary" and therefore will be used excessively if reimbursed, and sometimes on the grounds that they do not need to be covered by insurance because they are relatively inexpensive and hence affordable. The range of services to be provided is therefore highly relevant in primary care policy, especially for services provided directly by primary care practitioners.

Comprehensiveness can be achieved either by having a range of services available that are directly provided when the need arises, or by arranging for services to be provided elsewhere. In small facilities, such as the office of an individual practitioner, the range of available services is necessarily narrower than in larger facilities, because efficiency dictates that equipment and facilities be available only if use is sufficiently high. In these small facilities, with a narrow range of services, the provider must ensure that an appropriate referral is made and that the needed services will be received.

Comprehensiveness of care is an important mechanism because it ensures that services can be tailored to health needs. When services are too narrow in scope or depth, preventable illnesses may not be prevented, illnesses may be more severe than they need to be, quality of life may be jeopardized, and people may die sooner than they should. Patients themselves recognize the importance of comprehensiveness and express dissatisfaction in its absence. For example, Fletcher and colleagues (1983) found comprehensiveness to be the second most important characteristic of care (after continuity) for patients attending a primary care clinic for adults. Hickson and colleagues (1988) found that the third most common reason for parents' dissatisfaction, after lack of response to treatment and inconvenient location of office, was the failure of physicians to be interested in the behavior problem of their children.

> I went to see an ENT doctor because of pain in my ear. He poked in my ear, poked it this way and that way and he said that he didn't see anything wrong. I said "But doctor, it hurts me so bad that it even hurts me in my eye." He looked in my ear again with that light and he poked around and then he said that it's all in my imagination. When he said that it's all in my imagination I started to get angry, and when he said that, I stood up and said "Thank you very much for not finding anything wrong with my ear, Doctor. How much is your fee?" He told me and I said "Why don't you imagine that I just paid you?" And I walked out.
> They get angry with you and it makes you want to start screaming "Hey, I'm having the pain. Why don't you listen to me for five minutes? I come to you for help and you tell me I'm not having those pains and its all in my mind." I don't feel they should tell me anything until they listen to me first and then come to a conclusion of what's wrong. If there isn't anything wrong with my ear, then the doctor should find out what is giving me the pains and not assume that, just because I don't have a problem in the part of the body they take care of, that there's nothing wrong. I like doctors, but they have to listen and have a little compassion for the patient and not just think about their little piece of my body.   Mrs. R, a sixty-year-old secretary

The Benefits of Comprehensiveness

Few studies document the magnitude of benefit from comprehensive care. The benefits of comprehensive care can be inferred from the known benefits of preventive care (Russell, 1986), and from the benefits of providing services known to be effective for populations with particular health needs. One cross-national study, however, takes issue with the assumption that all needed services should be provided or arranged for by the source of primary care. After comparing the health systems in the United States with those in several western European nations, Silver (1978) concluded that children were more likely to have received indicated preventive care such as immunizations in systems where responsibility for these preventive procedures was divorced from the primary care practice and assumed by a public authority. Whether this is the case generally or just when primary care services are not adequate remains to be determined.

Policy and Research Implications of Comprehensiveness of Care: The Benefit Package

From the viewpoint of policymaking, decisions about the range of services to be provided or to be covered by insurance policies are critically important.

Insurance policies have always been written with specified benefits, but the standards for determining which benefits are included are generally not explicit.

With increasing mandates for public accountability, a corollary of expanded public financing for services, the composition of benefit packages is likely to come under greater scrutiny.

In the best of all possible worlds, the range of services provided would be determined only by need. In our world, compromises are necessary because resources are limited and there are competing needs for various types of health services and for social services other than health, such as housing and education.

In any health system, health care organization, or office practice, decisions must be made concerning the services to be directly provided or arranged for. Although not all services will be provided in the primary care sector, the primary care provider bears the responsibility for recognizing health care needs and arranging and coordinating services to be provided elsewhere. Therefore, the nature of the benefit package has high salience in primary care, even if it does not provide all needed services.

Sometimes certain services are mandated because of the requirement to provide them as a condition for reimbursement for services. More often, other considerations dictate the services that are to be available. These considerations are of several types: the availability of standards, the specification of needs to be addressed, values and priorities, and the marketplace. The resulting policy depends on which approach is taken. A discussion of each of these four considerations follows.

Standards

Prototypes of standards for benefit packages are available from professional associations. The Health Policy Agenda is the product of several years of deliberation by representatives of the major professional associations in the United States. In general, its package of specified benefits includes "physician services both inpatient and outpatient, hospital services, laboratory and roentgenogram services, prescription drugs, institutional care for the elderly and physically or mentally disabled, dental services, early and periodic screening including diagnosis and treatment services, family planning services, home health and personal care services, and other medically necessary services" (Tallon, 1989).

More explicit is the policy on health benefits proposed in 1989 by the American Academy of Pediatrics: "Health insurance plans should cover the following services for children from birth through 21 years old. These services should be ordered by a doctor and delivered in either a hospital or doctor's office . . . . the following is an ideal list of benefits . . . by which each plan's covered services should be compared." The list includes the following:

- Medical care, including health supervision and treatment of acute conditions and diagnosis of severe and long-term illness.
- Treatment of preexisting conditions.
- Surgical care, including steps to correct birth defects.
- Mental health, alcohol, and other substance abuse services.
- Emergency and trauma services.
- Inpatient hospital services.

- Consultations with pediatric subspecialists.
- Family planning services.
- Pregnancy services including prenatal care, prenatal consultation with a pediatrician, and care for the pregnancy of a single dependent of the policyholder.
- Care of all newborn infants including a pediatric specialist to attend high-risk pregnancies.
- Exams and health checks from the time of birth, and treatment of birth defects and other illnesses or injuries.
- Laboratory and pathology services.
- Diagnostic and therapeutic radiology services; anesthesia services.
- Services ordered by the doctor to help a child recover from an illness or injury including physical therapy.
- Speech therapy and occupational therapy.
- Home health care or the services of a licensed private duty nurse.
- Hospice care for the terminally ill child.
- Help for parents or guardians who need relief from the constant demand of caring for children with long-term illnesses.
- Long-term care services delivered in an intermediate care or skilled nursing center instead of in hospitals.
- Case management and supervision of the special medical and social services needed by chronically ill or disabled children.
- Medical and social services to check for and treat suspected child abuse or neglect.
- Transfer for transport to the hospital or health center.
- Preventive and restorative dental care.
- Nutrition counseling and checkups.
- Prescription drugs, medical and surgical supplies.
- Corrective eyeglasses or lenses.
- Hearing aids and special nutritional supplements.
- Rental or purchase of durable medical equipment, such as special beds or crutches.

The list of services, extensive as it is, still excludes certain services such as home visits by physicians and is vague about the extent of other services. Does surgical care include cosmetic surgery other than correction of birth defects? What is included under laboratory services? Nonetheless, it is obvious that the proposal of the American Academy of Pediatrics is far more comprehensive than most existing insurance benefits or public programs.

Standards for preventive care may be derived from published guidelines for a wide variety of types of procedures including screening, counseling, and immunizations or chemoprophylaxis. These guidelines (U.S. Preventive Series Task Force, 1989) review the effectiveness of more than one hundred clinical interventions for the prevention of sixty diseases and conditions, making recommendations concerning their indications and the timing and frequency of the services that were judged to be effective by a large panel of experts who reviewed the evidence.

Another approach to the specification of standards for providing services involves the development of goals and objectives for health status. In the United States, the Surgeon General of the Public Health Services developed such goals and objectives for the year 1990 and for the year 2000; the goals refer to the achievement of community health levels for a variety of conditions. It is possible that these objectives could set priorities for the range of services provided by health insurers or health care organizations (USDHHS, 1980; USDHHS, 1990).

Community Health Needs

The second aspect of benefit packages is the specification of the community's range and needs. In one method, a health care organization documents the health needs of its population, using the data to set priorities. In contrast to setting priorities according to uniform goals and objectives, as just described, this method could lead to different priorities between communities and depends on what data each community decides to collect and the criteria used to interpret it.

Evidence of Effectiveness of Services

The third element of the benefit package sets priorities according to the degree of benefit afforded. Instead of using data to set community health needs, this approach establishes priorities according to the public's perception of value of an intervention, its impact on outcome of care, and its cost.

This method has been attempted in Oregon, where a governmental commission has the responsibility for obtaining the needed information and developing the range of services that will be made available. The commission seeks public input through town meetings and random surveys. People are asked how highly they value such conditions as mobility, mental capacity, freedom from pain, and ability to continue working. The data then become part of a cost–benefit formula involving consideration of the extent to which a specific intervention achieves the desired outcomes. Priorities are set for each age group separately, to avoid competition of needs between different age groups, such as the elderly versus children (Welch, 1989).

Although the priorities are intended only for recipients of Medicaid in Oregon, other insurers will likely find the approach and its results useful. For example, proposed legislation in the U.S. Congress (Shortell and McNerney, 1990) will mandate basic benefits for all health insurance; what eventually is included in these bills may result only from perceived political expediency, from an Oregon-like approach, or from a combination of both. Data from the National Health Interview Survey, the National Ambulatory Medical Care Survey, the Health Care Expenditure Survey, and ongoing data collected by the Health Care Financing Administration can be drawn on and supplemented by research on the impact of medical care on health status and outcomes, to provide more refined estimates for priority setting in resource allocation (Shortell and McNerney, 1990).

The Marketplace

In a market-based health system, facilities often use comprehensiveness of services to gain an edge over competitors. A baseline of comprehensiveness is often determined by governmental regulations that mandate services such as office-based medical care; hospitalization; heart, kidney, or cornea transplants; and infertility control. Most additions to these basic packages derive from negotiations with employers or insurance companies, who often put their requirements out for bids by different health providers. The final negotiated benefit package is often revised based on experience and subsequent annual negotiations.

This final approach to the benefits package considers financial circumstances. In the United States, comprehensiveness of benefits is generally based on conditions in the marketplace; that is, to be competitive, insurers vie with one another to offer a wider variety of benefits, sometimes at considerably increased costs. As a result, the benefits available to individuals, families, enrolled groups, or employees in various industries or firms are widely discrepant. A comparison of fifteen insurance plans, each developed to provide financing for the previously uninsured, found that some services were always covered, some were usually covered (in ten to twelve plans), some were covered only sometimes (six to eight plans), and some were covered only rarely (one to three of the plans). The following were *always* covered:

- Visits to doctor's office.
- Outpatient diagnostic x-ray and laboratory testing.
- Outpatient surgery including doctor and facility charge.
- Well baby care.
- Ambulance services.
- Emergency room services.
- Hospital inpatient services including semiprivate room and board, surgeon's fees, anesthesiologist's fees, doctor visits in the hospital, and prescriptions.

However, not all plans covered these services to the same extent. Those services that were *almost always* covered included outpatient routine physicals, outpatient immunizations, outpatient physical therapy, and private duty nursing in the hospital, although again, the extent of the benefit varied.

Services that were *usually* covered included outpatient prescriptions, home health visits, and routine hearing and eye examinations.

Services that were covered only *some of the time* included convalescent care or care in a skilled nursing facility, outpatient mental health services, inpatient mental health services, and hospice care. Typically, these benefits were limited in the number of such services or the dollar amount provided.

Services that were provided only *rarely* included durable medical equipment, prosthetic and orthotic appliances, podiatry, and genetic testing and counseling (Alpha Center, 1990).

Up to the present, decisions about the content of the benefit package have largely been determined by professional opinion and cost considerations. Where needed services are not included, they are either not received at all, purchased by those who need them, or provided directly by the public health sector. Providing services through the public health sector runs the risk of fragmenting services and reducing the likelihood of achieving longitudinality, first contact care, and coordination of care. In developing nations, the focus on selective rather than integrated care may be justified, given the stage of underdevelopment of the health services sector and the type of population needs (Walsh and Warren, 1979). But even in this situation, a case can be made that integrated care is a more appropriate strategy in the long run (Rifkin and Walt, 1986).

In the future, decisions concerning the appropriate range of services to be provided in primary care, as well as secondary and tertiary care, will be based increasingly on public accountability, taking into consideration community needs, equity of access, and the effectiveness of services. Although methods for setting priorities are still primitive and there is no consensus concerning the approach to be used, increasing recognition of the limits on resources will focus attention on the challenge and compel careful consideration of the various alternatives.

Measurement of Comprehensiveness

As with the other attributes of primary care, the assessment of comprehensiveness can focus either at the system/population level or at the facilities/patients level. At the population level the needs of the community define the range of services that should be provided, and comprehensiveness is measured by the degree to which these services penetrate the community. At the facilities level, the range of services is usually set by professional guidelines such as immunization schedules or by recommendations for procedures such as mammograms. Assessment at this level generally uses chart audits or other records such as claims forms.

Range of Services

At the *population level,* standards for the range of services needed by communities are stated in such general terms that measurement of a range is generally not feasible. For example, the World Health Organization (1978) recommends that

> in order for primary care to be comprehensive, all development-oriented activities should be interrelated and balanced so as to focus on problems of the highest priority as mutually perceived by the community and health system, and that culturally acceptable, technically appropriate, manageable, and appropriately selected interventions should be implemented in combinations that meet local needs. This implies that single-purpose programs should be integrated into primary health care activities as quickly and smoothly as possible.

This statement makes it clear that comprehensiveness does not consist of single-purpose programs, but it gives little guidance to determine which problems are of highest priority and how decisions are to be made concerning the measurement of cultural acceptability, technical appropriateness, and manageability.

Therefore the methods for specifying a needed range of services are not clear. However, comparisons between various systems of care might provide information to establish relative standards. For example, a population-based approach to assessing "range of services" might entail comparing programs between countries, to determine the extent to which various types of services are covered by insurance (e.g., mental health services, preventive health services and screening, or dental health services). Few studies of this type exist, and a notable one is the International Collaboration Study of Medical Care Utilization (Kohn and White, 1976). One part of this study described selected aspects of health care provisions in the late 1960s in seven study countries: Canada, the United Sates, Argentina, the United Kingdom, Finland, Poland, and Yugoslavia. (Kohn and White, 1976, p. 420–37)

A *facilities-based* approach to assessing range of services compares the benefit packages of various types of health plans or health care organizations and the office equipment available to handle a broad range of services. In addition, assessment should consider the competence of the staff to deal with a broad range of problems, or the existence of a systematic approach to obtaining the services elsewhere if they are not provided in the facility.

Measurement of Needs Recognition

At the *population level,* comprehensiveness is reflected by the extent to which the health services address evident needs of the population.

One approach suited to the assessment of comprehensiveness of care at the population level was developed by investigators at the Indian Health Services program in Arizona (Schorr and Nutting, 1977). The approach was concerned with how adequately the health system addressed certain recognized needs of the population and traced the care through its successive stages. The investigators identified seven health problems (hypertension, adult-onset diabetes, tuberculosis, urinary tract infection, head lacerations, infant gastroenteritis, and acute otitis media) and determined the adequacy of case finding (problem recognition) for the conditions, along with diagnostic procedures, therapy, and follow-up.

This approach to assessing comprehensiveness is particularly well-suited to situations in which the needs of the community are known from prior studies of mortality and morbidity.

An alternative approach to judging the adequacy of the range of needed services in a community assumes approximately equivalent health in different communities, particularly for conditions that are responsive to medical care. If services are adequately comprehensive, parity should exist across communities in the frequency and severity of those illnesses where preventing or treating of illnesses is effective.

The potential of this method to identify areas where improvement in comprehensiveness is needed is demonstrated by two studies, in the United States and in Canada.

In the U.S. study, Woolhandler and colleagues (1985) compared death rates at different ages among whites and blacks for conditions in which death can be prevented by medical care to rates for all causes of death. At all ages below sixty, the ratio of black to white preventable death rates was greater than the ratio of black to white deaths from all causes, indicating that at least some of the excess deaths were due to a failure of the health system. The disparity in death rates associated with preventable causes was particularly striking for youths and young adults. If the problem were simply greater susceptibility to life-threatening illness among blacks, the ratio for preventable conditions would be equal to the ratio for all causes. Unequal ratios suggest inadequacy of health services, resulting from disparities in access or in the breadth of services available to the two types of populations.

In Canada, Wilkins and Adams (1983) used data from vital statistics and health surveys in the five regions, differing in both population size and income levels, to calculate life expectancy and quality-adjusted life expectancy in males versus females. There were systematic differences in both measures of health status in areas with different family incomes. Because the Canadian health system is designed to reduce financial barriers to access to services, the differences are not likely to be due to differences in access; rather, they are likely to be due at least in part to services insufficiently broad to meet the health needs of the less affluent population.

At the *facilities level* one empirical approach to measurement depends on the availability of information on the types of problems and diagnoses that are seen and managed in different facilities and by different providers. In the absence of differences in population needs, there should be general parity in the frequency of conditions seen by different practitioners or by different groups of practitioners who purport to provide comprehensive care. For example, facilities whose rates of diagnosis of adult depression or childhood behavior problems differ are likely to differ in their comprehensiveness of care, unless it can be shown that the actual frequency of these conditions in their populations differed.

Data from the National Ambulatory Medical Care Survey (see Chapter 10) provide information on problems and diagnoses in a representative sample of office-based practices of different specialists in the United States. The data can be used to compare the distribution of problems and diagnoses in the practice of various primary care specialties, to determine if their comprehensiveness differs. If data are obtained from all types of facilities in a community or country, the results are tantamount to a population-based method for comparing comprehensiveness across types of physicians. Thus, even though it is national in scope, the National Ambulatory Medical Care Survey, which obtains information only from office-based practices, is technically not completely population based.

In a normative approach to assessing comprehensiveness, a standard is set and facilities are measured against this standard. The Institute of Medicine (1978) checklist for primary care includes seven questions in its section on comprehensiveness. Four of these addressed the availability of a range of services but can be reframed to capture the actual provision of services by determining if the facilities accomplish the activity. These questions are as follows:

> Is the practice unit willing to handle, without referral, the great majority (over 90 percent) of the problems arising in the population served (for example, general complaints such as fever or fatigue, minor trauma, sore throat, cough, and chest pain)?
> Are the practitioners in the unit willing, if appropriate, to admit and care for patients in hospitals?
> Are the practitioners in the unit willing to admit and care for patients in nursing homes or convalescent homes?
> Are the practitioners in the unit willing, if appropriate, to visit the patient at home?

Following are the three questions that directly address the process aspect of comprehensiveness:

> Are appropriate primary and secondary preventive measures used for those people at risk, such as immunization for tetanus or polio, early detection of hypertension, and control of risk factors for coronary disease?

Are patients encouraged and assisted in providing for their own care and in participating in their own health care plan such as instruction in nutrition, diet, exercise, accident prevention, family planning, and adolescent problems?

Do the practitioners in the unit support those agencies that promote community health, such as health education programs for the public, disease detection programs, school health and sports medicine programs, and emergency care training?

Although this approach was designed for use at the facilities level, it can also be adapted to services organized at the population level. For example, do all practices in an area handle the great majority of problems in the population? Are appropriate primary and secondary preventive services used for people at risk?

A third approach to measuring comprehensiveness is based on the extent of recognition of the patients' needs. Repeated studies have shown that practitioners often fail to recognize problems that patients express when they seek care. The deficiencies are particularly pronounced when the problems are psychological or psychosocial. To assess the extent of recognition of patients' needs, there has to be a mechanism to independently ascertain the nature of problems presented by patients. One technique is to have patients write their complaints and concerns while waiting to see the practitioner, or to have the receptionist or equivalent person question the patient at his or her appearance at the facility. These complaints are then compared with those identified by the practitioners on a special data collection form or written in the medical record.

What Is Known About the Attainment of Comprehensiveness

A neurologist I know is frequently asked by patients whether he will make sure their heart exam is alright, "while he's at it." He replies, "Certainly, if you remind me where your heart is."

DR. K, a pediatric resident

Comprehensiveness has been assessed in two ways. In some studies, the availability of a *range* of services was ascertained while others determined the *extent* to which needed services were provided.

Six examples of assessment are listed here. The first is based on the availability of range of services, while the remaining five use the extent to which services are provided. All the studies use the empirical rather than the normative approach.

1. In a facilities-based study of a representative sample of practitioners in Baltimore, Maryland, Weiner determined the proportion of practices that were able to provide their regular patients a breadth of services. These services were sigmoidoscopy, electrocardiogram, audiometry, pelvic exam, superficial biopsy, minor suturing, tonometry, pulmonary function tests, blood hematocrit/hemoglobin determinations, counseling of more than forty-five minutes, immunizations, and complete physical examination. A comprehensiveness index was created from six of the preceding measures. Internists, pediatricians, and family practitioners ranked highest, or most comprehensive, and psychiatrists and surgical subspecialists ranked lowest; with other types of specialists intermediate. The relatively high

variability for medical subspecialists indicated the great variability in achievement of comprehensiveness by these types of physicians (Weiner, 1981, pp. 127 and 171).

2. National data derived from logs kept by physicians (as in the National Ambulatory Medical Care Survey) have been used to assess the range of services provided by various types of physicians.

One such study (Starfield et al., 1984) ascertained the distribution of all visits made by children, including teenagers, according to the type of physician. It revealed the following:

- While about one-third of all outpatient visits by children were to pediatricians, only about one in eight visits for minor surgery were to them.
- Family physicians and general practitioners, on the other hand, saw approximately the same proportion of child visits for minor surgery, suggesting their care is more comprehensive, at least with regard to the performance of minor surgery.
- Pediatricians also were underrepresented in their dealing with psychosocial problems; they provided less than one-fifth of the care for such problems, although they were the regular source of care for more than one-third of children.
- Family physicians and generalists, on the other hand, provided approximately the same proportion of care for psychological problems of children as for those for whom they were the regular source.

These findings suggest that generalists may provide greater comprehensiveness of care directly to children than pediatricians, who focus relatively heavily on preventive care and on care for medical illnesses. Additional confirmation of the belief that family practice provides greater comprehensiveness derives from the finding that both family physicians and general practitioners saw relatively more young people for obstetric problems, as well as for environmental and economic problems.

3. A facilities-based study of a national sample of recent graduates of residencies in internal medicine and family practice now in office-based practice used questionnaires and logs of all their encounters with patients for three days. Family physicians saw relatively more patients with sprains and strains and with psychosocial problems. They saw more than three times as many adult patients with trauma, and many more patients for prenatal and postnatal visits than internists. The latter saw more patients for chronic illnesses, especially for young and middle-aged patients (Cherkin et al., 1986). These findings suggest that family physicians provide a greater variety of types of care than general internists, who focus more on managing of chronic medical illness.

4. A facilities-based study conducted in five types of facilities in different geographic areas found widely discrepant rates of diagnosis of certain types of problems among children (Starfield et al., 1980). Rates for psychosocial problems varied from 5 to 15 percent, with no consistency by type of practice. The variability between the two multispecialty group practices was as great as that between multispecialty groups, the pediatric group, and the family practice group. However, the rates of diagnosis of psychosomatic conditions, all of which had a somatic manifestation, was highly consistent (at about 9 percent). These findings suggest that

comprehensiveness, at least as manifested by the extent to which facilities dealt with psychosocial problems in children, varied across facilities probably as a result of factors other than the type of organization or primary care specialty.

A study of adults in several types of health care facilities in three U.S. cities reached similar conclusions. A standard diagnostic test for depressive disorder was given to patients, and the results were compared with visit report forms completed by the clinicians who saw the patients for the problem under care. Only about half of all depressed patients seen by their internist were diagnosed as depressed; the proportion varied from 46 to 51 percent, depending on the type of facility (Wells et al., 1989).

5. In a study comparing prepaid group practice with fee-for-service practice, 1,580 adult and children were randomly assigned to receive care free of charge from either a fee-for-service physician of their choice or a prepaid group practice. Patients receiving their care from the prepaid group practice had higher rates of preventive care than patients receiving care from fee-for-service physicians. This study indicated that comprehensiveness, at least in preventive care, can differ according to the type of organization (Manning et al., 1984).

6. A facilities-based study of the extent to which professional guidelines for preventive procedures were met found no consistent difference between generalists (family physicians and general internists) and subspecialty internists. However, patients who had a complete physical exam were more likely to have the indicated procedures. The most salient determinant was the extent to which the physician believed the procedure to be important; patients whose physicians believed the procedure to be important were more likely to have them performed. This study indicates the importance of individual professional practice patterns in comprehensiveness of care in situations with no particular mechanisms within the facility to ensure the performance of indicated procedures (Dietrich and Goldberg, 1984).

Adherence to guidelines for comprehensiveness cannot be assumed. For this reason some medical facilities have adopted computerized systems to remind physicians that certain procedures are indicated for individual patients. To further improve the performance of these procedures, stamped postcards addressed to the patients may be given to the physicians with the reminders. Better adherence to the guidelines results when these mechanisms are in place (Schoenbaum, 1990).

Summary

Comprehensiveness of services requires an explicit scope of services and recognition of situations in which their application is appropriate. Thus, it should be assessed by examining the range of services available (a structure feature) and recognition of needs (a process feature).

Practice- or facilities-based studies can provide information concerning the extent to which a full range of services is available and even provided to patients who receive care within it, but it cannot provide the information necessary to determine whether the facility meets the needs of the population for which it assumes responsibility.

Assessment of the adequacy of comprehensiveness is compromised in the absence of an accepted terminology for specifying needs. When no well-validated standards exist for specifying the range of services, population differences in the frequency of conditions amenable to medical care serve to identify inadequate comprehensiveness.

Although primary care physicians generally provide more comprehensive care than specialists, disparities in comprehensiveness persist among the primary care specialties, and even within them.

The policy imperative related to comprehensiveness concerns the nature of the benefit package provided by various insurers and health services organizations. At present, these benefits vary widely. In the future, approaches employing considerations of population needs, effectiveness of various types of interventions on health status and health outcomes, and costs of the interventions will provide a more consistent and rational approach to the definition of the range of services to be made available.

References

Alpha Center, Key Features of HCUP Products. Program Update, No. 10, Washington D.C., June 1990, p. 9.

American Academy of Pediatrics. Guidelines for Your Family's Health Insurance. Elk Grove Village, Ill., 1989.

Cherkin D, Rosenblatt R, Hart L, and Schleiter M. A comparison of the patients and practices of recent graduates of family practice and general internal medicine residency programs. Med Care 1986; 24:1136–50.

Dietrich A and Goldberg H. Preventive content of adult primary care: do generalists and subspecialists differ? Am J Public Health, 1984; 74:223–27.

Fletcher R, O'Malley M, Earp J, Littleton T, Fietcher S, Greganti A, Davidson R, and Taylor J. Patients' priorities for medical care. Medical Care 1983; 21:234–42.

Hickson G, Stewart D, Altemeier W, and Perrin J. First steps in obtaining child health care: selecting a physician. Pediatrics 1988; 81:333–38.

Institute of Medicine. A Manpower Policy for Primary Health Care: A Report of a Study. Washington D.C., National Academy of Sciences, 1978.

Kohn R and White KL. Health Care: An International Study. London, Oxford University Press, 1976.

Manning W, Leibowitz A, Goldberg G, Rogers W, and Newhouse J. A controlled trial of the effect of a prepaid group practice on use of services. N Engl J Med 1984; 310:1505–10.

Report of the U.S. Preventive Services Task Force. Guide to Clinical Preventive Services. Baltimore, Williams & Wilkins, 1989.

Rifkin S and Walt G. Why health improves: defining the issues concerning "comprehensive primary health care" and "selective primary health care." Soc Sci Med 1986; 23: 559–66.

Russell L. Is Prevention Better than Cure? Washington D.C., The Brookings Institution, 1986.

Schoenbaum S. Implementation of Preventive Services in an HMO Practice. J Gen Intern Med 1990; 5(suppl) S123-27.

Schorr G and Nutting P. A population-based assessment of the continuity of ambulatory care. Medical Care 1977; 15:455-64.

Shortell SM and McNerney WJ. Criteria and guidelines for reforming the U.S. health care system. N Engl J Med 1990; 332(7):463-67.

Silver G. Child Health: America's Future. Germantown, Md., Aspen Systems Corp., 1978.

Starfield B, Gross E, Wood M, Pantell R, Allen C, Gordon B, Moffatt P, Drachman R, and Katz H. Psychosocial and psychosomatic diagnosis in primary care of children. Pediatrics 1980; 66:159–67.

Starfield B, Hoekelman R, McCormick M, Benson P, Mendenhall R, Moynihan C, and Radecki S. Who provides health care to children and adolescents in the United States? Pediatrics 1984; 74:991–97.

Tallon JR, Jr. A health policy agenda for including the poor. JAMA 1989; 261:1044.

U.S. Department of Health and Human Services (USDHHS). Public Health Service. Promoting Health/Preventing Disease. Objectives for the Nation. Washington D.C., fall 1980.

U.S. Department of Health and Human Services (USDHHS). Public Health Services. Promoting Health/Preventing Disease: Year 2000 Objectives for the Nation. Washington D.C., 1990; DHHS Publication No. (PHS)90-50212.

Walsh J and Warren K. Selective primary health care: an interim strategy for disease control in developing countries. N Engl J Med 1979; 301:967–74.

Weiner J. The Baltimore City Primary Care Study: An analysis of office-based primary care. Unpublished thesis, The Johns Hopkins University School of Hygiene & Public Health, 1981.

Weiner J and Starfield B. Measurement of the primary care roles of office-based physicians. Am J Public Health 1982; 73:666–71.

Welch HG. Health care tickets for the uninsured: First Class, Coach, or Standby? N Engl J Med 1989; 321(18):1261–64.

Wells K, Hays R, Burnam A, Rogers W, Greenfield S, and Ware J. Detection of depressive disorder for patients receiving prepaid or fee-for-service care: results from the medical outcomes study. JAMA 1989; 262:3298–302.

White KL. Life and death and medicine. Scientific American 1973; 229:23–33.

Wilkins R and Adams O. Health expectancy in Canada, late 1970s: demographic, regional, and social dimensions. Am J Public Health 1983; 73:1073–80.

Woolhandler S, Himmelstein D, Silber R, Bader M, Harnly M, and Jones A. Medical care and mortality: racial differences in preventable deaths. Intl J Health Services 1985; 15:1–22.

World Health Organization. Primary Health Care. Geneva, 1978, p. 25.

6

Coordination and the Processes
of Referral

I've been to so many doctors in the last few months, I need a doctor to put it all
together.

A patient in an emergency room, 1990

Coordination, the fourth component of care, is essential for the attainment of each
of the other features. Without it, longitudinality would lose much of its potential,
comprehensiveness would be difficult, and the first contact function would become
purely administrative. Descriptions of primary care from the physician's vantage
often refer to the primary care professional as the patient's advocate (Robinson,
1977) or in terms of the primary care physician's commitment to people (McWhinney, 1975; Draper and Smits, 1975). To accomplish what these terms imply, the
primary practitioner must be aware of all of the patient's problems in whatever
context they arise, at least insofar as they relate to health.

Coordination, according to the dictionary, is a "state of being harmonized in a
common action or effort." This definition formally expresses what more down-to-
earth descriptions imply. The essence of coordination is the availability of information about prior problems and services and the recognition of that information as it
bears on needs for current care.

The term *case management,* which was recently used to describe cost-containment efforts, may overlap with the notion of coordination of care, but there is little
agreement on a standard operational definition of case management. This book
considers case management to be a function of care deriving from the presence of a
regular source, that is longitudinality (see Chapter 4).

The following account illustrates the challenges of coordination:

> A 41-year-old woman is under care in a "primary care clinic" for mild hyperten-
> sion. This vignette contains information about her care from one visit in the clinic
> to her next scheduled follow-up, a period of approximately three months. For
> purposes of assessing coordination, interest is in the extent to which the primary
> care practitioner recognized information relevant to the woman's care that arose in
> the interval between the initial visit and the follow-up visit.
>
> - At her "index" visit in the primary care clinic, the physician noted that
>   patient had a "cold," and had signs consistent with this diagnosis. The

physician ordered a chest x-ray and symptomatic therapy. At follow-up three months later, the practitioner made reference to the respiratory infection in his notes, but made no mention of the results of the x-ray and whether the patient had taken the medication or felt any relief from it.

- Two weeks after the index visit the patient appeared without appointment in the same clinic still complaining of symptoms of the respiratory infection. A prescription for an antibiotic was given but the primary care practitioner made no reference to it in his follow-up note.
- Two weeks later the patient appeared unexpectedly in another clinic in the same facility, with the same type of symptoms; the practitioner noted that she took valium regularly and smoked. He recommended discontinuing smoking and additional symptomatic therapy.
- At the follow-up visit in the primary care facility, the practitioner made no mention of this visit or the therapy or advice in his notes in the medical record.
- One month later (and one month prior to the scheduled primary care follow-up visit), the patient was seen in the emergency room after an apparent suicide attempt. Again, there was no reference to this visit or to the associated problem in the primary care physician's note at follow-up.

Review of the patient's medical record for the period prior to the index visit revealed that she had been suffering from headaches for several years and was being followed in the neurology clinic, where no abnormalities were found on repeated tests. At one visit, mild hypertension was noted and she was referred to the primary care clinic, where she had been seen every three months to follow the progress of the mild hypertension. On at least one prior occasion, one year earlier, she had attempted suicide through an overdose of her medication. Despite this history, the patient's serious mental health problems were not integrated into her primary care. There was little or no coordination of care for this patient by the primary care physician. (Starfield et al., 1977)

This case report indicates the types of problems for which coordination is critical.

Benefits of Coordination

This section addresses the benefits of continuity specifically as it relates to coordination of care.

Several studies indicate that recognition of information about patients can be enhanced by improving the mechanisms of continuity—for example, continuity of practitioner results in better recognition of patient information. In a series of studies conducted in six different primary care facilities, continuity of practitioner improved the extent of recognition of various types of information, particularly of clearly identified problems and therapies. Continuity of practitioner did not, however, facilitate practitioner recognition of visits elsewhere (Starfield et al., 1976; Simborg et al., 1978).

The inclusion of *problem lists* to medical records results in improved problem recognition. Studies in the six clinics cited previously indicate that such improvements in medical records facilitated recognition of information. Problem lists improved recognition of patients' problems, especially if the duration between visits was long and a different practitioner was seen at each one (Simborg et al., 1976). Similar benefits accrued from the insertion of computerized printouts of information. This resulted in improved physicians' recognition about patients' previous problems and therapies, but it did not improve recognition of visits made elsewhere between one visit to the primary care facility and a subsequent follow-up. The improvements occurred whether or not there was continuity of practitioner from one visit to the next in the primary care facility (Starfield et al., 1977).

Computerized feedback of certain types of information also enhances patient care by facilitating coordination. Improved continuity in the form of a computerized summary in the medical records in a clinic lacking continuity of physicians can improve various aspects of care. One study consisted of a randomized controlled trial involving patients who were followed in three medical specialty clinics. A summary containing updated information about the patient included a problem list, medications, results of laboratory tests, and suggested actions concerning the process of care. Those patients whose physicians were given the summaries were more likely to have indicated procedures and referrals actually carried out, greater implementation of indicated diets, and more discovery of both new and resolved problems. These patients also spent fewer days in the hospital, on average (Rogers and Haring, 1979).

Improved continuity through a computerized profile of drugs can also result in more effective use of information. In one study, the computer generated an updated summary of each patient's current and past medications whenever a drug was dispensed from the pharmacy. Using a random sample of patients seen in an outpatient setting, this profile was inserted within twenty-four hours into their medical records. The duration of drug interactions experienced by patients was shorter for these patients than for those without the recorded profiles. The profiles were also associated with a reduced number of visits over a year of study, presumably because of a decrease in the number of visits for drug prescription renewals.

Patients are more satisfied with their care when they can see the same practitioner in successive visits. In a study conducted in eleven southern California ambulatory health care settings, Linn (1975) assessed satisfaction both in general and with the specific physician, and found large differences across the facilities, most of which could be attributed to differences in continuity of physician.

The recognition of important information by practitioners can also be improved by various other methods. For example, the presence of an observer during the practitioner–patient interaction can heighten a practitioner's sense of accountability to the patient so that the practitioner may inquire about problems of importance to the patient. One study demonstrating this effect showed that the presence of an observer enhances the recognition of problems the patients thought required follow-up (Starfield et al., 1979a).

In studies such as those summarized, the effect is probably a result of improved recognition of patients' needs by practitioners, and therefore of greater agreement between patients and physicians concerning the patient's problems. When patients and practitioners agree on the patient's problems, the problems are more likely to improve than when only the patient or only the physician notes their existence. This has been found to be the case whether improvement in the patient's health was judged by the physician (Starfield et al., 1981) or by the patient (Starfield et al., 1979b).

Policy and Research Implications of Coordination of Care: The Referral Process

The management of illness is becoming increasingly complex. As technology advances, there is greater possibility for diagnosis, intervention, and more specialization of personnel to provide the new services. The nature of illness is also changing: the aging of the population and the increased survival of individuals with complex and disabling conditions places an ever-increasing burden on the health care system for coordination of care. Support of the coordination function depends on the development of better means of information transfer, especially when patients receive care from a variety of sources to which they have been referred.

Since referrals require patients to see different practitioners, continuity cannot be achieved by having the same practitioner; other mechanisms are needed. Historically, medical records have provided this mechanism, but standard medical records are not adequate for the task. Most of them serve solely as memory aids and are useful primarily to the practitioner who wrote them; other practitioners are much less likely to use the information even if available.

Some improvements in medical record systems are designed to facilitate other purposes, such as billing and reimbursement, documenting the process of care, reviewing quality, monitoring, or legal functions. They do not improve recognition of information from referral visits or those made elsewhere on the patient's initiative.

Early studies of the referral process and associated information transfer were conducted in England. After the 1948 reorganization, the hospital outpatient departments became the center to which a general practitioner could refer a patient for a specialist's opinion. Several studies were conducted in individual hospital clinics to determine the type of patients practitioners had referred. The findings of these studies (Chamberlain, 1965) were as follows:

- Almost 10 percent of letters from practitioners accompanying patients were less than fifteen words long; only 15 percent had more than one hundred words.
- There was a marked association between the letter's length and social class for people who lived near the hospital. The higher the social class, the longer the letter.
- Fewer than half the letters requested that the patient see a specific consultant. The largest proportion of referrals contained no tentative diagnosis but merely requested that the consultant "see and advise."
- The quality of the communication between doctor and consultant was poorer for doctors who lived closest to the hospital.

In the 1950s, the now classic studies of Williams and his colleagues in North Carolina examined the process of referral and the characteristics of information associated with them. In these studies, referring physicians and their referred patients were interviewed, and the medical records were reviewed. Adequacy of the referral process was judged according to the following five criteria:

- Definition and specification of the need and purpose of referral, including mutual understanding between patients and physicians.
- Adequate communication of purpose to consultant.
- Attention to purpose by consultant.
- Adequate communication of findings and recommendations to the referring physician.
- Clear understanding by patient, referring physician, and consultant of responsibility for the patient's continuing care (Williams et al., 1961).

Specification of the purpose of the referral was determined for 99 percent of the referrals. The purpose was specified only a quarter of the time; a nonspecific medical reason was given in another one-third of referrals. The most common reason (39 percent of the time) was "in response to a patient's request." About one in twenty-five referrals was related to the patient's inability to afford medically needed services.

The reason for the requested consultation was specified in a communication in only four of ten instances; it was more likely to be found in the written communication when the physician had stated it clearly in the interviews.

In only one of six instances was the reason for referral mentioned in the communication from the consultant back to the general practitioner.

Criteria concerning communication back to the referring physician was a policy of the medical center and was achieved 100 percent of the time. However, subsequent communications were never sent for 24 percent of patients for whom they were warranted.

Responsibility for continuing care of the patient was studied by examining the medical record and interviewing the patients and referring physicians as to whether the patient had returned. Patients were more likely to return to the referring physician when the medical record clearly specified such an intent. In total, 62 percent of patients returned to their general practitioner, but the percentage was 79 percent if the note was clear and only 42 percent if the note was unclear as to the disposition (Williams et al., 1961).

Recent studies in Great Britain confirm these earlier findings. A substantial proportion of referral letters omit such information as details of prior drug therapy and illnesses. The adoption of fixed format letters has not been widespread, even though they more consistently contain needed information. Letters from consultants to primary care physicians are often unresponsive or at best unclear in aiding the primary care physician and are often considerably delayed (Wilkin and Dornan, 1990). General practitioners, patients, and consultants have different perceptions of the reason for referral, and these differences are generally not explicit. In such situations, it is difficult, if not impossible, to judge the appropriateness of the referral, the nature of the information transferred, and the procedures for the referral and the return to the primary care physician (Grace and Armstrong, 1987).

Giving the patient *written* information may be very important. In a survey of three hundred consecutive consultations carried out in England, both patients and physicians were asked what advice they were given about the need for a return visit.

Physicians and patients agreed less than half the time about what advice was given. Patients were especially likely to disagree with their physician when no specific advice was given during the physician–patient encounter; in these instances, patients were much more likely to indicate that a return visit was necessary (Armstrong et al., 1990).

One study, conducted in a major HMO in the United States, confirmed the importance of written information; if a written form was given for an appointment to return to the referring physician, the patient was much more likely to return to that physician than if the information was provided verbally. Also, patients were more likely to return to their physician if they considered the physician to be their family doctor (Lawrence and Dorsey, 1976).

Despite the high salience of the referral process from the viewpoint of both increasing need for referrals and the high costs of associated care (Glenn, Lawler, and Hoerl, 1987), few more recent U.S. studies concern information transfer and coordination of care. One such study showed that consultants communicate their findings to referring practitioners in only 55 percent of the consultations. If the patient was given a return appointment to the primary care physician, the physician was more likely to know the result of the consultation (McPhee et al., 1984).

Unfortunately, the absence of a clear distinction between secondary (consultative) care and tertiary (referral) care impedes the development of information systems tailored to specific needs. At least one study has shown that secondary clinics can be distinguished from tertiary clinics; both continuity of practitioner and coordination of care within the facility are higher in clinics characterized as tertiary than in clinics characterized as secondary. Despite these differences, specialists working in these clinics were unable to specify whether patients were under their care only for short-term advice and guidance or they perceived a long-term responsibility for more comprehensive care than specialists generally provide (Mawajdeh, 1988).

An understanding of the relative responsibilities of primary care physicians and specialists can facilitate care of patients, as was demonstrated in a study in Great Britain of problem drinkers referred to a special alcohol clinic by their general practitioners. After an initial session, in which they received advice and counseling in the clinic, the patients were randomly allocated to either a group that received continued care from the clinic or a group that returned to their general practitioner, who was contacted and supported by the specialists from the clinic. After six months, patients cared for by their general practitioner had improved at least as much as if not more than those cared for by specialists. Those seen by specialists made many more visits to the clinic, whereas those cared for by general practitioners made more visits to the more convenient general practitioner's offices. In this case, clearly defined responsibility for short-term care by specialists with continuing advice to the primary care physician improved the efficiency without any sacrifice in effectiveness of care (Drummond et al., 1990).

The challenge to development of policy regarding coordination of care depends on obtaining more knowledge about the process of referral, the transfer of information deriving from referrals, and the extent to which primary care physicians recognize this information. Policy issues include the development of at least some standards for referral that can be taught during the process of medical education and used for monitoring the quality of various components of medical care. New methods of information transfer are required to facilitate the recognition of information generated in visits to physicians other than the primary care physician. The suggestion that client-held records, or "medical passports," may be a fruitful approach has been made by at least one major commission, The National Commission to Prevent Infant Mortality (1988). It is also likely that better coordination will require more explicit recognition at the health system level of the relative roles and responsibilities of primary, secondary, and tertiary care.

The coordination between these three sectors is known as regionalization. Under regionalization, linkage between primary care, secondary care, and tertiary care is well-established. Services are organized according to the needs of the population and designed to include enough practitioners and facilities at each level to take care of, but not exceed, the needs at that level.

The need for resources at each level can be calculated either empirically, on the basis of the experience of existing health systems, or by estimating the number of required practitioners and facilities at each level from theoretical calculations based on existing needs or demands.

In many countries, care is organized to provide one primary care practitioner for each 1,000 to 3,000 patients, depending on the age and illness level of the population. Community hospitals serve as a resource for consultations and for uncomplicated hospitalizations. Tertiary care hospitals are generally affiliated with medical teaching institutions and serve as a place to refer patients with unusually complicated or rare problems.

Theoretical calculations of the need for resources at each level can be based on professional judgments about what services should ideally be provided. The "needs-based" assessment was used by Lee and Jones (1933) and subsequently by a group at Yale University (Schoenfeld et al., 1972) in the early 1970s. A panel of experts developed estimates of the average number of visits required to render primary care for specific acute and chronic conditions for different age groups.

In the Yale estimates, each acute condition was judged to require one and one-half visits per year, on average; nonelderly adults with chronic conditions and those over age sixty-five with chronic conditions were judged to require three and seven visits per year, respectively. Data from the National Health Interview Survey (see Chapter 10) were used to determine the prevalence of acute and chronic conditions in the different age groups. The next step in the process of estimation required judgments concerning how many visits the average physician could handle. From these estimates the number of physicians required was calculated.

Practice-based surveys (such as the National Ambulatory Medical Care Survey—see Chapter 10) provide data on the number and type of visits that are managed by physicians so that professional estimates are no longer required for this step in the calculations. The Graduate Medical Education National Advisory Committee (GMENAC) provided greater precision for the estimates made by the earlier efforts. Just substituting the actual visit rate for the theoretical visit rate increased the estimate for the number of required physicians by 20 to 40 percent, depending on the specialty.

Estimates based on demand for services use information such as that in the National Health Interview Survey, in which respondents are asked about the number of visits they have made to physicians. Corrections can be made for anticipated changes that will influence utilizations (such as expanded health insurance), or for changes in the anticipated balance between nonphysician and physician providers. Using these types of estimates, GMENAC estimated the number of physicians needed in both primary care and other physician specialities. In general, demand-based estimates are considerably lower than need-based estimates.

As is the case for the facilities level, any process of estimating the need for primary care physicians relative to need for specialist physicians requires much more attention to the roles of primary care physicians and what they are trained to manage rather than refer.

Measurement of Coordination

The achievement of coordination on the population level may be assessed by aggregating measures at the individual level. As is the case with each of the other three attributes, the assessment of coordination at the individual level requires consideration of a structural element of care, which provides the potential for the achievement of the attribute, and a process feature, which provides the behavioral counterpart.

The structural element essential to coordination is continuity of care; there must be a mechanism that makes the practitioner aware of problems wherever they arise, so that possible interrelationships can be detected and managed. Continuity is generally achieved by having the same physician or other health professional care for the patient from one encounter to the next. But there are other mechanisms for achieving continuity, such as a team of individuals whose communication channels permit them to convey important information about patients to one another; a medical record containing accurate and complete information about the patient; a computer containing information about the patient, or even records carried by patients themselves.

The paramount behavioral element in coordination is the process of problem recognition or recognition of information about the care of problems. Only when problems are recognized can practitioners act on them. Theoretically, coordination requires such awareness, but since no treatment can be given without the initial recognition, this step is critical. When physicians do not act on information, it constitutes an error of omission rather than one of commission; that is, the problem arises from failure to recognize the existence of information rather than failure to act on it.

Measuring Continuity

To achieve coordination of care, there must be a mechanism to transfer information about a patient's problems or the care received for these problems. Continuity, in the context of measuring coordination, involves the presence of such a mechanism to

TABLE 6.1. Summary of Longitudinality versus Continuity versus First Contact

	Longitudinality	*Continuity*	*First Contact*
Type of Feature	Structure/Process	Structure	Structure/Process
Measured by	Utilization over time of the regular source of care	Continuity meausre	Accessibility to the regular source and use of that source for each patient-initiated visit
Problem-oriented?	No. Essence is use of the regular source of care regardless of the nature or type of problem, in which a personal relationship is established and maintained.	Yes. A mechanism to provide information follow-up of problems or types of problems	No. Most problems would be expected to be new, recurrences of old problems, and often undifferentiated
Personal relationship required?	Yes	No. Could be achieved by other means	No
Time specific?	Yes—a relationship over time	No. Relates to information generated in a series of two or more visits	No. Specific to a particular event in time
Person oriented?	Yes	No. Problem oriented	Yes
Shorthand summary	Over time—personal relationship	Episodes of illness (or chronic illness)	Single event
Suggested measures	UPC (Breslau & Reeb) COC (Bice & Boxerman) LICON (Steinwachs) K Index (Ejlertsson & Berg)	SECON (Steinwachs) LISECON (Steinwachs) 'S' Index (Smedby et al., 1984)	Population or patient interview

ensure an uninterrupted succession of events between visits. Longitudinality and continuity differ in that, in the former, the succession of events is time-bound and across the full spectrum of potential problems or reasons for visits. For continuity, the important issue is the succession of events between visits, regardless of where they occur or why. Many studies on the subject of continuity have been published, but most fail to distinguish between the separate concepts of longitudinality (the presence and use of a regular source of care over time) and continuity (the sequence of visits in which there is a mechanism of information transfer). Since the use of the same term to connote different phenomena impedes the research process and accumulation of knowledge, it is important to distinguish between the concepts of longitudinality and continuity (Starfield, 1980). Table 6.1 compares these concepts and, in addition, distinguishes them from first contact.

Patients often lack an identifiable regular source of care yet achieve continuity of care for particular problems. That is, it is possible to have continuity without longitudinality. For example, a complete medical record may be available to all providers while patients see a different provider or go to a different facility for each visit. The situation includes continuity but no longitudinality.

Conversely, patients may have a regular source of care in which longitudinality is reasonably well achieved, yet lack continuity for events that occur in sequence. For example, a patient who always sees the same provider achieves longitudinality, but there may be no mechanism, apart from human recollection, for information transfer. Furthermore, continuity of care can be a feature of tertiary care as well as primary care, in the sense that individuals who are followed by tertiary care specialists for their rare or complex conditions can achieve a high degree of continuity. For example, there was no difference between continuity of care by generalists and by subspecialists for patients with chronic illness seen in a study of middle-aged and elderly adults in Quebec (Beland, 1989) or in one concerning patients in large, private multispeciality practices in California (Goldberg and Dietrich, 1985).

There are several methods of assessing continuity of practitioner and others for evaluating continuity achieved by alternative means.

GINI and CON (Standardized Index of Concentration) (Shortell, 1976). These two measures of continuity have their roots in international economics, where they are used to assess the concentration of resources. They are only useful when applied at the population level. Shortell applied them to assess the source of utilization during an episode of illness involving at least five visits. The measures cannot, however, distinguish patterns of use; seeing one physician for several visits and another physician for other visits would result in the same index as patients who alternated physicians for the total number of visits. Thus, the measures cannot capture the succession of events within a sequence of care.

Sequential Continuity (SECON). This measure is the fraction of sequential pairs of visits with the same provider. It ranges from zero to 1, with an expected value of 1 divided by the number of available practitioners if care-seeking and scheduling are random (Steinwachs, 1979).

Likelihood of Sequential Continuity (LISECON). This measure is the likelihood that SECON (sequential continuity) is greater than would occur if providers were distributed randomly across sequential visits (Steinwachs, 1979).

Both SECON and LISECON are probability measures that control for the level of use and the number of practitioners available.

"S" index. In this method (Smedby et al., 1986), the patient is identified at a particular visit. This measure considers whether the provider seen on that occasion is the same provider seen at the previous visit to the facility. It takes a value of 1 if the provider is the same and a value of 0 if not. Values can be aggregated by averaging over the visits of each person.

Closed-Loop Referral Rate. (Holmes et al., 1978). This method measures the availability of information concerning referrals. It is based on information in medical records and ascertains the percentage of referrals that resulted in a return of information about the referral.

All of these measures are empirical, that is, relative measures. Because they do not yield standards for adequate continuity, their use is primarily for comparison between one facility and another or between one population group and another. The measure developed by the Institute of Medicine, on the other hand, is normative, that is, it has a definable standard against which measurements can be graded.

The IOM Measures. The Institute of Medicine (1978) checklist contains three questions concerning continuity. Can a patient who desires to do so make subsequent appointments with the same provider? Are complete records maintained in a form that is readily retrievable and accessible? Are relevant items or problems in the patient's record highlighted, regularly reviewed, and used in planning care?

Of the six measure of continuity, only the last two address mechanisms of continuity other than that of continuity of personnel. Two of the IOM criteria explicitly recognize the importance of medical records. Their role in achieving continuity has been demonstrated by several studies. For example, Martin (1965) showed that patients are more likely to keep appointments when the follow-up dates are written in the medical record. Another study (Zuckerman et al., 1975) showed that patients are more likely to have information about their care (for example, drug dosage and drug actions) if the information is in the record.

For records to serve as a mechanism of continuity, they must contain important information about patients. There are no standards for medical records, for computerized information systems, or for records kept by patients themselves, except for the most general categories including diagnosis and therapy. Not all types of information are equally well recorded in medical records. For example, Osborne and Thompson (1976) conducted an extensive nationwide study of pediatric practitioners and showed that many items of information were widely considered as evidence of high-quality care but were not recorded. Items that were consistently recorded included certain routine measurements in infancy and childhood, symptom history in children with an acute recurrent condition, drug therapy, and a program for continuing treatment in children with allergy, chief complaints, symptoms, as well as some types of diagnostic tests. The approach to assessing the adequacy of medical records for purposes of achieving continuity compares actual occurrence with what was recorded and is ascertainable. This involves several techniques:

• Interviewing patients and physicians.

To obtain information by this method, both patients and physicians should be queried, because they do not always agree on the events related to the visit. In fact, studies have shown that (1) patients and their physicians agree on the nature of the patient's problems in only half of all visits; (2) in one-quarter of the visits, a problem is mentioned only by the patient; and (3) in one-quarter of the visits a problem is mentioned only by the physician (Starfield et al., 1979; Romm and Putnam, 1981; Starfield et al., 1981).

• Audiotaping and videotaping the encounter (Zuckerman et al., 1975).

• Having an observer record the content of a visit and comparing it with the medical record (Starfield et al., 1979).

Innovations in medical records can improve their usefulness for purposes of continuity. Problem lists inserted in the front of the record do improve practitioners' recognition of problems under care. Computerized summaries also achieve this purpose for some but not all aspects of care. Highlighting of information, either in a handwritten note (Williamson et al., 1967) or by computer (Barnett, 1976) achieves the same purpose. (Chapter 10 supplies more detail on information systems.)

Measuring Recognition of Information Requiring Follow-up

Four methods have been used to assess the recognition of information concerning patients' problems or the care provided for them.

IOM measures (Institute of Medicine, 1978). Answers to several questions serve as a normative measure of the achievement of coordination. Each deals in some way with the transfer of information to facilitate integration of care, although none actually assess the recognition of the information. The questions are as follows: (a) Do the practitioners in the unit furnish pertinent information to other providers serving the patient, actively seek relevant feedback from consultants and other providers, and serve as the patient's ombudsman in contacts with other providers? (b) Is a summary or abstract of the patient's record provided to other physicians when needed? (c) Do the practitioners in the unit develop a treatment plan that considers the ability of the patient to understand it? (d) Do the practitioners use a variety of tactics to ensure that the patient will cooperate in the treatment? (e) Does the plan of treatment reflect the patient's physical, emotional, and financial ability to carry it out? The first two questions facilitate coordination by improving the likelihood that practitioners will recognize important information. The last three questions are more related to the quality of care than to the process of coordination itself.

The University of North Carolina method (Fletcher et al., 1984) assesses coordination more directly. Coordination is defined as written evidence that the specialist is aware of the primary care physician's involvement with the patient and that the primary physician either arranged the visit or knew about it beforehand. Alternatively, coordination is achieved if the primary physician was aware of a visit that a patient already made to another physician.

The Dartmouth Primary Care Cooperative Project (COOP), a collaborative practice-based network in northern New England, further refined the assessment of coordination of care. Care was considered to be coordinated only if the primary care physician was aware in advance of visits elsewhere. This could occur either when the primary care physician arranged for a visit elsewhere, when the physician discussed the need for it before the patient made the appointment, or when the physician knew in advance of visits that patients planned elsewhere on their own initiative. Information about visits was obtained from periodic phone calls to patients who kept diaries recording all of their contacts with physicians, hospitalizations, nonphysician ambulatory visits, and filling of drug prescriptions. Coordination was present for 75 percent of ambulatory physician visits, 81 percent of nonemergency hospitalizations, and 78 percent of all drug prescriptions, but for only 33 percent of visits to nonphysician providers (Dietrich et al., 1988).

In another approach to assessing coordination, patients are asked where and for what reason they have gone for care. Providers are asked if the patient has been seen elsewhere and, if so, where and for what. Alternately, information systems may be used as the source of information on where patients went and for what reason. This alternative is useful primarily where the information system contains data on all visits made by patients; if claims forms are obtained for all visits, the data base containing them would be a suitable source of data. Recognition of the information by providers is assessed either by querying them as to their knowledge of these visits and their content or by reviewing their records for evidence of recognition of the information. This is the method used in the study that provided the case history earlier in this chapter (Starfield et al., 1977).

Certain aspects of information recognition can be obtained directly from the medical record. For example, is there a mechanism for providers to sign off on laboratory results, and do they conform? Are there mechanisms to ensure that patients know they need a follow-up appointment or procedure, and are these mechanisms implemented and recorded?

What Is Known About the Attainment of Coordination

Many primary care as well as specialty care facilities lack high levels of continuity. Studies described here provide evidence of this.

In a study conducted over a period of seven months in six different facilities (three pediatric, three adult), there was wide variability in the percentage of patients seen by the same practitioner on two successive visits in which the second was a follow-up for the first. Percentages in five facilities ranged from 46 to 70 percent and was above 90 percent (93 percent) in only one. Despite the fact that approximately one in nine patients in each facility made a visit elsewhere on recommendation of their own practitioner and even more made a visit on their own, medical records of these patients often did not contain information about these visits (Table 6.2; Starfield et al., 1976).

Smedby et al. (1984) calculated sequential continuity rates in visits for nine different problems in a primary care facility. The percentage of visits with continuity of practitioner ranged from 10 percent in the case of obstructive lung disease to 28 percent for diabetes. Continuity rates were higher for scheduled visits than for unscheduled ones and for those in which the patient was seen by a physician who had served at the facility for longer periods of time. Overall, continuity was only 22 percent.

Holmes et al. (1978) compared the closed-loop referral rate for patients of family physicians who had a family practice residency with patients of general practitioners whose knowledge was derived only from their practice experience. The two types of physicians did not differ in their closed-loop referral rate, which was about 80 percent.

TABLE 6.2.   Continuity of Care by Medical Records:
Percentage of Charts Containing Content of Intervening Visits

	Suburban Group		Urban Group		Hospital Clinics	
Scheduled visits						
Adults	96%	(25)	92%	(13)	87%	(39)
Pediatric	77%	(22)	100%	(10)	82%	(33)
Unscheduled visits						
Adult	81%	(164)	63%	(96)	32%	(244)
Pediatric	83%	(100)	36%	(39)	54%	(39)

Source: Starfield, Barbara, et al. Studies on Continuity and Coordination of Care. The Johns Hopkins University, 1979.

There are no known comparisons of the actual attainment of coordination by different types of physicians, since most studies of coordination were done within particular institutions.

In one study of the extent of coordination between the internal medicine primary care facility and other facilities at the University of North Carolina, the percentage of visits that were coordinated ranged from 22 percent for specialty clinics to 47 percent for "walk-in" visits (Fletcher et al., 1984). Coordination for visits to the emergency room was 28 percent. In contrast, the percentage of hospital admissions for which there was coordination with the primary care facility was 75 percent. Overall, only about one-third (35 percent) of instances requiring coordination achieved it.

A second study of the attainment of coordination was conducted in six different primary care facilities—three internal medicine and three pediatric—all affiliated with a large medical center. It involved patients who were given an appointment for

TABLE 6.3.   Effect of Provider Continuity on Recognition
of Information

Type of Information	Provider Continuity	Percentage of Instances in which Information Was Recognized
Clearly identified problems	Yes	75.3
	No	62.3
'Problem-in-text'	Yes	58.2
	No	52.7
Therapies	Yes	52.2
	No	43.0
Tests done	Yes	31.0
	No	22.7
Scheduled intervening visits	Yes	47.2
	No	41.8
Unscheduled intervening		17.3

Source: Adapted from Starfield et al. Medical Care 1976; 630.

a follow-up visit within six months of a previous visit and who were identified at the follow-up visit. Before their second visit, their records were reviewed to ascertain the information generated in the first visit as well as that generated in subsequent visits elsewhere, either on referral by the primary care facility or self-initiated. In all facilities, the recognition of information generated within the primary care clinic, whether or not the patient was seen by the same practitioner as in the initially identified visit, was much greater than the extent of recognition of information generated in visits made elsewhere. Table 6.3 shows this to be the case for all types of information: (1) clearly identified problems, that is, those highlighted in the medical record because they were recorded as chief complaints or diagnoses; (2) problems mentioned in the text of medical record notes; (3) therapies prescribed; or (4) tests done (Starfield et al., 1976). The extent of achievement of coordination for visits occurring elsewhere was similar to that obtained in the University of North Carolina study.

Summary

Achieving coordination is a challenge for primary care practitioners. Patients frequently make visits elsewhere, either on the recommendation of their physician or at their own discretion, and medical records frequently do not contain information about these visits or what occurred in them.

Continuity of practitioner facilitates recognition of information concerning the care of patients within the primary care facility but does not improve recognition of information about visits elsewhere.

Currently available improvements in medical records facilitate the recognition of information generated about patients and improve the care provided to patients, but they do not extend to recognition of information obtained elsewhere, either on referral or on the patient's initiative.

The greater the recognition of patient's problems by practitioners, the more likely the patient is to show subsequent improvement.

These generalizations are derived from a relatively small number of studies in only a few facilities. Informed policy decisions concerning coordination of care require more research on referrals and the information associated with them, and the employment of measures to clearly distinguish between the concepts of continuity and coordination in the context of both consultative (secondary) and referral (tertiary) care.

References

Armstrong D, Glanville T, Bailey E, and O'Keefe G. Doctor-initiated consultations: a study of communication between general practitioners and patients about the need for reattendance. Br J Gen Pract 1990; 40:241–42.

Barnett GO. Computer-Stored Ambulatory Record (COSTAR). NCHSR Research Digest Series. DHEW Publication No. (HRA) 76-3145. US Department of Health, Education, and Welfare, 1976.

Béland F. A descriptive study of continuity of care as an element in the process of ambulatory medical care utilization. Canad J Public Health 1989; 80:249–54.

Bice T and Boxerman S. A quantitative measure of continuity of care. Medical Care 1977; 15:347–49.

Breslau N and Reeb K. Continuity of care in a university-based practice. J Med Educ 1975; 50:965–69.

Chamberlin J, Acheson R, Butterfield W, and Blancy R. The population served by the outpatient department of a London teaching hospital. A study of Guy's. Medical Care 1965; 4:81–88.

Dietrich A, Nelson E, Kirk J, Zubkoff M, and O'Conner G. Do primary physicians actually manage their patients fee-for-service care? JAMA 1988; 259:3145–49.

Draper and Smits W. The primary care practitioner—specialist or jack-of-all-trades. N Engl J Med 1975; 293:903–7.

Drummond D, Thom B, Brown C, Edwards G, and Mullan M. Specialist versus general practitioner treatment of problem drinkers. Lancet 1990; 336:915–18.

Ejlertson G and Berg S. Continuity of care measures. An analytic and empirical comparison. Med Care 1984; 22:231–39.

Fletcher R, O'Malley M, Fletcher S, Earl J, and Alexander J. Measuring the continuity and coordination of medical care in a system involving multiple providers. Medical Care 1984; 22:403–11.

Glenn JK, Lawler FH, and Hoerl MS. Physician referrals in a competitive environment: An estimate of the medical impact of a referral. JAMA 1987; 258:1920–23.

Goldberg H and Dietrich A. The continuity of care provided to primary care patients: a comparison of family physicians, general internists, and medical subspecialists. Medical Care 1985; 23:63–73.

Grace J and Armstrong D. Referral to hospital: perceptions of patients, general practitioners and consultants about necessity and suitability of referral. Family Practice 1987; 4:170–75.

Graduate Medical Education National Advisory Committee. GMENAC Staff Paper #1. Physician Manpower Requirements. DHEW Publication No. (HRA) 78-10. 1978. Rockville, Md. U.S. Department of Health, Education, and Welfare. Public Health Service, Bureau of Health Manpower.

Holmes C, Kane R, Ford M, and Fowler J. Toward the measurement of primary care. Milbank Mem Fund Q 1978; 56:231–52.

Institute of Medicine. A Manpower Policy for Primary Health Care: A Report of a Study. Washington D.C., National Academy of Sciences, 1978.

Lawrence R and Dorsey J. The generalist–specialist relationship and the art of consultation. In: Noble J. (ed). Primary Care and the Practice of Medicine. Boston, Little Brown & Company, 1976.

Lee R and Jones L. The Fundamentals of Good Medical Care. Chicago, University of Chicago Press, 1933.

Linn L. Factors associated with patient evaluation of health care. Milbank Mem Fund Q Fall 1975; 531–48.

Martin D. The disposition of patients from a consultant general medical clinic. In: White KL (ed). Medical Care Research. Oxford, Pergamon Press, 1965.

Mawajdeh S. Levels of care in pediatric specialty clinics. Unpublished doctoral disseration, The Johns Hopkins University School of Hygiene and Public Health, Baltimore, 1988.

McPhee S, Lo B, Saika G, and Meltzer R. How good is communication between primary care physicians and subspeciality consultants? Arch Int Med 1984; 144:1265–68.

McWhinney I. Family medicine in perspective. N Engl J Med 1975; 293:176–81.

National Commission to Prevent Infant Mortality. Death Before Life: The Tragedy of Infant Mortality. Washington D.C., August 1988.

Robinson D. Primary medical practice in the United Kingdom and the United States. N Engl J Med 1977; 297:188–93.

Rogers JL and Haring OM. The impact of a computerized medical record summary system on incidence and length of hospitalization. Medical Care 1979; 17:618–30.

Romm F and Putnam S. The validity of the medical record. Medical Care 1981; 19:310–15.

Schonfeld H, Heston J, and Falk I. Numbers of physicians required for primary medical care. N Engl J Med 1972; 286:571–76.

Shortell S. Continuity of medical care: conceptualization and measurement. Medical Care 1976; 14:377–91.

Simborg D, Starfield B, Horn S, and Yourtee S. Information factors affecting problem follow-up in ambulatory care. Medical Care 1976; 14:848–56.

Simborg D, Starfield B, and Horn S. Physician and nonphysician health practitioners: the characteristics of their practices and their relationships. Am J Public Health 1978; 68:44–48.

Smedby B, Smedby O, Eriksson E, Mattsson L-G, and Lindgren A. Continuity of care. An application of visit-based measures. Medical Care 1984; 22:676–80.

Smedby O, Eklund G, Eriksson E, and Smedby B. Measures of continuity of care: a register-based correlation study. Medical Care 1986; 24:511–18.

Starfield B. Continuous confusion? Am J Pub Health 1980; 70:117–19.

Starfield B, Simborg D, Horn S, and Yourtee S. Continuity and coordination in primary care: their achievement and utility. Medical Care 1976; 14:625–36.

Starfield B, Simborg D, Johns C, and Horn S. Coordination of care and its relationship to continuity and medical records. Medical Care 1977; 15:929–38.

Starfield B, Steinwachs D, Morris I, Bause G, Siebert S, and Westin C. Presence of Observers at Patient Practitioner Interactions: Impact on Coordination of Care and Methodologic Implications. Am J Public Health 1979a; 69:1021–25.

Starfield B, Steinwachs D, Morris I, Bause G, Siebert S, and Westin C. Patient-provider agreement about problems. Influence on outcome of care. JAMA 1979b; 242:344–46.

Starfield B, Wray C, Hess K, Gross R, Birk P, and D'Lugoff B. The influence of patient–practitioner agreement on outcome of care. Am J Public Health 1981; 71:127–32.

Steinwachs D. Measuring provider continuity in ambulatory care: An assessment of alternative approaches. Medical Care 1979; 17:551–65.

Steinwachs D, Weiner J, Shapiro S, Batalden P, Coltin K, and Wasserman F. A comparison of the requirements for primary care physicians in HMOs with projections made by the Graduate Medical Education National Advisory Committee. N Engl J Med 1986; 314:217–22.

Thompson H and Osborne C. Office records in the evaluation of quality of care. Medical Care 1976; 14:294–314.

Wilkin D and Dornan C. GP Referrals to Hospital: A review of research and its implication for policy and practice. Center for Primary Care Research, University of Manchester, July 1990.

Williams TF, White KL, Fleming WL, and Greenberg BA. The referral process in medical care and the university clinic's role. J Med Ed 1961; 36:899–907.

Williamson J, Alexander M, and Miller G. Continuing education and patient care research. JAMA 1967; 201:938–42.

Zuckerman A, Starfield B, Hochreiter C, and Kovasznay B. Validating the content of pediatric outpatient medical records by means of tape-recording doctor–patient encounters. Pediatrics 1975; 56:407–11.

III

POLICY FOR PRIMARY CARE

A

Primary Care in the United States

7

Characteristics of Practice
and Practitioners

Primary Care and Specialism: A Recent History

This chapter addresses two issues of high salience in the organization and delivery of primary care: the balance between specialization and "generalism," and the content of primary care. It begins with a history of the growth of specialization and proceeds to a discussion of the problems it has induced. Statistical information will provide material for analysis of the adequacy of medical manpower in meeting the challenges of primary care. The chapter concludes with a discussion of the present imbalance and its consequences.

Although the term *primary care* has a long history, it was virtually unknown in the United States before the mid-1960s. Even today, it is not widely adopted by the medical profession, which uses the terms family medicine, general internal medicine, and general pediatrics to reflect the concepts that are embodied in the broader term primary care. It seems likely that this is a result of the emphasis on specialization that has long characterized U.S. medicine.

Specialism was well underway during the second decade of the century (Stevens, 1978). In 1915, 66 percent of graduates of medical schools said they eventually planned to subspecialize. Ophthalmology was the first medical specialty to be formally organized (Wechsler, 1976). In the 1930s, many specialties emerged as separate entities; by 1937 there were formal certifying boards in ophthalmology (1917), otolaryngology (1924), obstetrics/gynecology (1927), dermatology (1932), pediatrics (1933), orthopedic surgery (1934), psychiatry and neurology (1934), radiology (1934), proctology (1935)—later to become colon and rectal surgery, urology (1935), internal medicine (1936), and surgery (1937). Subsequent formalization was achieved for neurological surgery (1940), physical medicine (1947), preventive medicine and public health (1948), and thoracic surgery (1950). Several of these specialties later changed their names, and the American Board of Thoracic Surgery was not officially recognized as a separate major specialty until 1970. Thus, by 1950, there were over fifteen boards to certify the competence of physicians entering specialty fields, and none for generalists (Wechsler, 1976).

The availability of federal funds to provide assistance for education of veterans of World War II accelerated the entry of physicians into specialties, and as a result the proportion of physicians who were generalists fell rapidly from 75 percent in 1935 to 45 percent in 1957 (Knowles, 1969). By the mid-1960s, a crisis in the availability of primary care physicians was widely anticipated, and several national commissions recommended steps to reverse the trend with proposals to establish a specialty of family medicine. However, the traditional specialists actively resisted the establishment of a specialty of general/family practice. In 1965 the American College of Physicians (internists) went only so far as to state their interest and feeling of responsibility to promote the family practice of medicine (Knowles, 1969). After repeated attempts of the Academy of General Practice to establish a certifying board in family practice, in 1969 one was finally approved. In 1971 Congress passed legislation (Health Professions Educational Assistance Program), which for the first time authorized the expenditure of funds to support training in family medicine. Since 1968, many states, starting with New York, have enacted laws calling for the development of family practice programs in state medical schools. By 1972, thirty-one schools had created departments of family practice and another thirty had set up divisions of family practice; these accounted for about three-fifths of all medical schools at the time. There were 107 approved residency programs in family medicine: 63 in community hospitals, 41 in university or university-affiliated hospitals, and 3 in military hospitals, with a total of more than 1,000 residency positions (Rousselot, 1973). *Family practitioner* is the designation for graduates of these approved family medicine training programs; *general practitioner* is generally reserved for similar physicians who graduated from medical school before these training programs were developed.

Legislation passed in 1963 and 1968 facilitated expansion in the overall supply of all physicians, including specialists, but in 1971 federal legislation to support the growth of family medicine failed to affect the supply of primary care physicians. Table 7.1 indicates that the surgical and other nonmedical subspecialists experienced greatest growth shortly after the 1963 legislation and that the number of medical specialists other than general internists grew most rapidly in the mid-1970s.

In 1976, concern about the maldistribution of physicians (relatively low physician/population ratios in most nonurban areas) and the lagging supply of primary care physicians led Congress to pass legislation that provided funding for primary care training programs in family medicine, general internal medicine, and pediatrics. The number of family physicians, internists, and pediatricians and the ratio of such physicians to the population increased, more than compensating for the decline in the number of general practitioners. The ratio of primary care physicians to the population increased from 69.6 in 1980 to 78.8 in 1985 and 83.7 in 1990. Despite this increase, the growth in number of primary care physicians has been less than that of non-primary care specialists so that they comprised a smaller proportion of all physicians in 1990 than in 1985.

Between 1978 and 1988, the growth rate in the supply of all active physicians was 30 percent, with wide variability depending on the specialty. The highest growth rates were in diagnostic radiology, gastroenterology, and pulmonary diseases, all having growth above 60 percent. Family medicine, internal medicine, and pediatrics grew at rates of 12, 37, and 29 percent, respectively (calculated from

TABLE 7.1. Number (per 100,000 population) of Professionally Active Physicians, Selected Years 1963 and Following[a]

| | 1963 | 1966 | 1968 | 1970 | 1972 | 1974 | 1975 | 1976 | 1978 | 1981 | 1982 | 1990[c] | Projected in 1985 | |
													1990	2000
Total active MDs[b]	134.8	141.8	144.0	148.3	150.7	152.7	156.1	158.6	169.2	184.5	190.6	226.3	223.4	245.3
Primary care	56.7	56.6	56.7	56.2	57.7	58.5	59.9	61.9	64.6	76.7	73.6	84.3	88.4	102.8
General/family practice	34.4	31.7	29.9	27.7	26.0	25.0	25.0	25.3	25.3	27.9	26.4	28.3	29.6	32.8
Internal medicine	15.7	17.5	18.7	20.0	22.5	23.9	24.9	26.4	28.2	34.9	33.9	39.5	41.5	48.7
Pediatrics	6.6	7.4	8.1	8.6	9.2	9.6	10.0	10.2	11.1	13.9	13.3	16.4	17.3	21.3
Other medical specialties	6.3	7.0	7.7	8.3	7.8	8.1	8.7	8.6	10.2	12.1	12.2	16.6	15.0	18.6
Surgical specialties	34.9	37.8	39.8	41.1	42.8	43.2	44.0	44.9	46.1	51.6	50.4	55.0	55.2	60.4
Other specialties	36.9	40.4	39.8	42.8	42.4	43.0	43.4	43.2	48.4	50.7	54.4	64.1	56.3	63.6

[a]In addition, there were 25,479 osteopathic physicians in the United States in 1986. Osteopathy is a separate and smaller branch of medicine that is based on a different philosophy of pathology than in the predominant allopathic medicine. Of all osteopathic physicians in 1986, 61% practiced primary care, mostly general practice. Osteopathic physicians comprise 3.9% of all physicians in the United States, but they represent 9.3% of primary care physicians and 12.2% of family physicians (Barnett and Midting, 1989).

[b]Excludes unknown specialties.

[c]Includes inactive physicians (1990 total only).

Sources: 1963–76—U.S. Department of Health, Education, and Welfare, 1978, pp. iv–46; 1978—Glandon and Shapiro, Statistical Abstract of the United States, for population estimates 1981, 1982, 1990—American Medical Association, 1984 and 1992; 1990 and 2000—Bureau of Health Manpower, p. 29.

94 POLICY FOR PRIMARY CARE

TABLE 7.2. Population per Physician:
United States, 1990*

Total active physicians	454
Primary care physicians	1,186
General/family physicians	3,529
Internists	2,529
Pediatricians	6,082 (1,758 children*)
Other medical specialists	6,009
Surgical specialists	1,819
Other specialists	1,559

Source: Calculated from Table 7.1.
*Under age 20.

Table 7.1). From a high in the 1980s, the percentage of medical students who expressed an intent to enter a primary specialty has declined from 14 to 7 percent in internal medicine, from 7 to 5 percent in pediatrics, and from 18 to 14 percent in family medicine. Furthermore, a substantial proportion of physicians in primary care actually practice with a subspecialty interest. In fact, 40 percent of all professionally active internists indicate that they devote themselves to allergy, cardiovascular disease, critical care medicine, diabetes, endocrinology, geriatrics, hematology, immunology, infectious diseases, neoplastic diseases, nephrology, nutrition, oncology, or rheumatology. About one-third of all first-year residents in internal medicine enter a subspecialty residency in their second year (Andersen et al., 1990). About one in eight (12 percent) pediatricians indicate that they practice adolescent medicine, neonatal–perinatal medicine, allergy, cardiology, pediatric endocrinology, pediatric hematology–oncology, or pediatric nephrology (Barnett and Midtling, 1989). Almost 30 percent of physicians who classify themselves as pediatricians actually either have a subspecialty concentration in their practices (17 percent) or practice a subspecialty exclusively (11 percent) (McCrindle, 1989).

Between half and two-thirds of all internists in training pursue subspecialty training (IOM report, 1989; Barnett and Midtling, 1989). This increase in subspecialization is occurring in osteopathic medicine as well; between 1980 and 1987, the number of subspecialty certificates in osteopathy doubled.

Table 7.2 shows the overall number of persons for each type of physician in the United States in 1990. There was one active physician for every 454 people, but fewer than half of these were primary care physicians; on average, there was one primary care physician for every 1,200 people. Medical subspecialists serve populations that are about five times greater than that of primary care physicians. The degree of imbalance is most striking when viewed as the imbalance in proportion of physicians being trained in each separate specialty. For example, in the late 1980s there was one neurology resident in training for every five residents in family medicine (Rowley et al., 1990).

As is shown in Chapter 15, this specialty orientation is relatively unique in Western industrialized nations. Moreover, the growth in the supply of all primary care physicians combined is less than that in the supply of specialists, thus predisposing to an increasingly specialized orientation to medical practice in this country. There is even a trend toward subspecialization among primary care physicians; a decreasing proportion of primary care physicians are generalists, and an increasing proportion are internists and pediatricians who themselves often practice with a subspecialty focus.

The Content of Primary Care in the United States

The major source of information about the nature of primary care in the United States is the National Ambulatory Medical Care Survey (NAMCS), a national survey of office-based practices. It was first conducted in 1974 and continued every year until the early 1980s, when it was reduced to administration once every five years. In 1990, it again was made annual.

More than 80 percent of ambulatory visits made by patients in the United States are to doctors offices or to organized clinics; about 15 percent are in hospital outpatient departments and fewer than 2 percent are in the home (Health United States, 1989, Table 64). Therefore, the National Ambulatory Care Survey reflects the nature of ambulatory care for the vast majority of people in the United States. it does not, however, provide information on consultative or referral care provided in hospitals, or on primary care delivered to people who live in socioeconomically deprived areas of cities where there are relatively few office-based physicians.

The sample frame for NAMCS is a list of physicians maintained by the American Medical Association, which obtains the names of all medical students when they graduate from medical school and periodically requests information from them by survey questionnaire, and by the American Osteopathic Association, which keeps a roster of osteopaths.*

The sample frames are stratified by specialty and by geographic area; about three thousand physicians are requested to complete forms on visits made to their offices for one week of the year. The following fifteen items of information are listed: patient's birth date, ethnicity, race, sex, presenting problem or complaint, whether the patient has been seen before, whether the patient was referred by another physician, the type of reason for the visit, diagnoses (up to three may be listed), diagnostic and therapeutic services, disposition, and the names of all medications administered or prescribed. Additional data may be elicited in particular areas to obtain information on issues related to current national interest. Information is obtained on approximately fifty thousand visits per year; about 75 percent of all physicians asked to participate actually do so (NCHS, 1983). The survey is conducted under the aegis of the National Center for Health Statistics.

Diagnoses are coded using the International Classification of Diseases–Clinical Modification and drugs are coded according to a special system devised by the agency. A unique feature of the NAMCS is the coding of presenting problems by

---

*The NAMCS data are the only routinely collected source of information that includes doctors of osteopathy (DOs) with medical doctors (MDs). Unless otherwise indicated, all other data relate only to MDs because comparable information about osteopaths is not available. In the NAMCS data, DOs are not separately distinguished, so that it is not possible to determine whether and in what way their practices differ from that of MDs.

TABLE 7.3.   Percentage of Office-Based Visits by Age and Type of Physician:
United States 1985

	General/Family Practice	Internal Medicine	Pediatrics	Other Specialty
<15	25.0	2.2	55.0	17.7
15–24	35.6	6.4	6.0	52.0
25–44	31.9	9.1	1.0	57.9
45–64	32.0	15.7	—	52.3
65+	29.1	22.1	—	48.8
All ages	30.5	11.6	11.4	46.5

means of a unique system in which problems are categorized by "modules" as follows: symptoms; diseases; diagnostic, screening, and prevention; treatment; injuries and adverse effects; test results; or administrative. The information obtained from NAMCS provides the basis for studies to illuminate the nature of primary care, to explore reasons for differences, and to assess the impact of differences on benefits and costs of care.

Table 7.3 shows the distribution of office-based visits by age of patient and by type of physician. General and family physicians provide about a third of all care, as well as for adults in each age group. Just over half of all visits made by children under age fifteen are to pediatricians, generalists provide care in a quarter of children's visits, and other nonprimary care physicians provide it in about 18 percent of visits. Internists provide relatively more care to older people, but still account for only 22 percent of visits for people over age sixty-five.

Table 7.4 describes the reasons for visits to primary care and nonprimary care specialists by children under age fifteen. The majority of visits, both to primary and to other types of physicians, are prompted by symptoms; visits to generalists are significantly more likely to be for symptoms than visits to pediatricians. Only one in twenty visits are made specifically for a designated disease, with the proportion being slightly but significantly greater for pediatricians than for generalists. Pedi-

TABLE 7.4.   Visits by Children Age 0 to 15 to Office-Based Physicians,
by Reason for Visit and Type of Physician: United States 1985

Reason for Visit	Percentage of All Visits[a]		
	General/Family Physician	Pediatrician	All Physicians
Symptoms	65.9[b]	56.9	58.9
Disease	4.4[b]	5.2	5.9
Diagnosis/screening/prevention	12.9[b]	29.9	21.4
Treatment	5.1	3.1	5.8
Injury and adverse effects	5.6	2.6	4.7
Administrative purpose	5.6	1.5	2.4

Source: National Ambulatory Medical Care Survey, 1985 (courtesy of James DeLozier).

[a]Difference between 100% and column totals includes all other reasons and unknown reasons.

[b]Differences between adjacent columns are statistically significant ($p < .05$).

TABLE 7.5. Visits by Adults to Office-Based Physicians by Reason for Visit
and Type of Physician: United States 1985

Age and Reason for Visit	Percentage of All Visits[a]		
	General/Family Physician	Internist	All Physicians
15–24			
Symptoms	59.1	75.0	52.0
Disease	4.3	c	5.1
Diagnosis/screening/prevention	17.3[b]	8.6	23.9
Treatment	5.6	c	8.0
Injury and adverse effects	8.1	c	7.1
Administrative purpose	4.3	3.0	2.5
25–44			
Symptoms	62.3	63.8	54.5
Disease	6.6	13.5	7.6
Diagnosis/screening/prevention	15.8[b]	9.2	20.7
Treatment	6.4	5.0	9.6
Injury and adverse effects	5.1	3.0	4.3
Administrative purpose	2.4	2.9	1.4
45–64			
Symptoms	57.3	54.1	55.3
Disease	11.2	19.6	14.3
Diagnosis/screening/prevention	17.2[b]	14.2	13.2
Treatment	7.1	6.2	10.9
Injury and adverse effects	4.3	1.7	3.1
Administrative purpose	1.3	1.1	.8
Test results[d]	2.8	1.8	1.1
65+			
Symptoms	54.9	51.9	51.4
Disease	12.8	19.4	17.1
Diagnosis/screening/prevention	19.3[b]	15.8	14.5
Treatment	7.4	8.4	11.8
Injury and adverse effects	3.0	1.6	2.2

Source: National Ambulatory Medical Care Survey, 1985 (courtesy of James DeLozier).

[a]Difference between 100% and column total includes all other reasons and unknown reasons.

[b]Differences between adjacent columns are statistically significant ($p < .05$). Differences between primary care specialists and other physicians cannot be tested because of different sampling ratios.

[c]Estimate unreliable because of the small sample size.

[d]Estimates for test results in other age groups and for administrative purposes at age 65+ are unreliable because of the small sample size.

atricians also provide significantly more care prompted by diagnostic, screening, or preventive purposes than generalists. General/family physicians provide relatively more care for injuries than do pediatricians, although the differences do not reach statistical significance because of the small numbers. Several other studies have shown similar findings. The reasons for visits to specialists other than primary care physicians do not differ in a major way from those for visits to general/family physicians or pediatricians.

Table 7.5 presents the reasons for visits made by adults. Visits prompted by a need for diagnosis, screening, or prevention are consistently and significantly less common in the practice of internists than in the practice of general/family physicians. Internists provide a consistently greater proportion of care for specific diagnoses than general or family physicians, although the differences are not statistically significant for the different age groups separately. General or family physicians provide relatively more care for injuries experienced by adults in all age groups than do internists. Visits to nonprimary care physicians (reflected in the differences between "all physicians" and the primary care physicians) are consistently more likely to be for specific treatment and less likely to be for administrative purposes such as need for a certificate to attend camp or to return to work.

In summary, visits to general or family physicians are more likely than those to pediatrician or internists to focus on patients' complaints and injuries. In contrast, internists are more likely to focus on specific diseases and pediatricians on prevention and screening than general or family physicians. For all types of primary care physicians, however, the majority of visits are prompted by symptoms and complaints; only a small minority are prompted by specific diagnoses. In general, neither striking nor consistent differences exist between specialists and primary care physicians.

Chapter 1 indicated that primary care practice should be distinguishable from specialty practice by a greater variety of types of presenting problems and a greater variety of diagnoses. Table 7.6 shows that this is only partly the case in the United States. The table shows the similarity of family medicine, internal medicine, and general surgery when the number of the most common problems required to reach 50 percent of visits is considered. The number is much larger than for other specialties, even pediatrics. It is only when specialties are arrayed by the percentage of

TABLE 7.6. Number of Most Frequent Presenting Problems Accounting for 50% of All Visits and the Percentage of All Presenting Problems Contributed by the Most Common Presenting Problems, by Specialty of Physician (United States, 1985)

	Numbering of Presenting Problems Accounting for 50% of All Visits	Percentage of All Visits Accounted for by the 50 Most Frequent Presenting Problems
Family/general practice	27	65.3
Internal medicine	21	69.3
Pediatrics	6	84.9*
Cardiovascular medicine	7	90.5
Dermatology	5	87.0
General surgery	22	71.0
Obstetrics/gynecology	3	91.7
Ophthalmology	3	99.0
Orthopedic surgery	8	92.9
Otolaryngology	7	93.2
Urology	10	92.0
Psychiatry	4	97.0
Neurology	9	90.0

Source: National Ambulatory Medical Care Survey, 1985 (courtesy of James DeLozier).

*66.1% excluding well-baby visits.

TABLE 7.7. Number of Most Frequent Diagnoses Accounting for 50% of All Visits and Percentage of All Diagnoses Contributed by the 50 Most Common Diagnoses, by Specialty of Physician: United States, 1985

Specialty	Number of the Most Common Diagnoses Accounting for 50% of All Visits		Percentage of Visits for the 50 Most Frequent Diagnoses
	With Routine Checkups Included	Without Routine Checkups Included	
Family/general practice	31	38	60.8
Internal medicine	26	28	64.6
Pediatrics	7	22	83.1*
Cardiovascular medicine	6	6	85.1
Dermatology	4	4	96.9
General surgery	32	38	61.2
Obstetrics/gynecology	5	29	89.5
Ophthalmology	4	4	97.2
Orthopedic surgery	14	14	87.4
Otolaryngology	9	9	88.5
Urology	7	9	94.3
Psychiatry	2	2	99.5
Neurology	11	11	83.5

Source: National Ambulatory Medical Care Survey (courtesy of James DeLozier).

*63.5% after eliminating well-infant or well child visits.

visits accounted for by the top fifty problems that pediatrics, as well as family medicine and internal medicine, are separated from the more traditional "specialties." But even here, general surgery resembles primary care rather than other specialties, and pediatrics does so only when well-baby visits are not included.

Table 7.7 presents a similar picture regarding the variety of diagnoses in primary care and specialty practice. The table shows that this common perception is only partly accurate. It shows the number of the most common diagnoses which, considering their frequencies, account for 50 percent of all visits to different types of physicians in the United States. If the most common diagnoses made by primary care practitioners are more varied, these practitioners encounter *more* diagnoses in a set percentage of visits (for example, 50 percent) than specialists do. For physicians such as generalists, internists, and general surgeons, at least twenty diagnoses must be included to reach 50 percent of visits. For physicians such as dermatologists, psychiatrists, obstetrician–gynecologists, and ophthalmologists, five or fewer most common diagnoses account for 50 percent of visits. For pediatricians, the six most common diagnoses account for 50 percent of the visits, making them more like specialists than generalists. However, when the diagnoses of "well-child" or "well-person" are not included, twenty-two diagnoses are required to reach 50 percent of the visits to pediatricians, making them more like generalists. When the percentage of visits that are required to encompass the 50 most frequent diagnoses are considered, surgeons resemble family physicians and internists rather than specialists and pediatricians resemble them only when well-infant or well-child visits are excluded.

Therefore, neither variety nor type of problem or diagnosis clearly distinguishes specialists such as generalists, internists, and pediatricians, who are thought to administer primary care, from those such as surgeons, who are not considered to be primary care providers. This may be true only in the United States, where most of the population has direct access to a specialist. This direct access to specialists may lead to specialty care resembling primary care. In this situation, it is not possible to use specific diagnoses or variety of diagnoses as the basis for identifying primary care practice, at least with current coding schemes that cannot distinguish between severities of the problem or diagnosis.

Primary care practices are assumed to have a larger percentage of visits classified as prevention-related but, as is the case for variety of diagnoses and problems, the data in the United States do not uniformly support this distinction. The percentage of adult visits that are unrelated to symptoms varies from 1.8 for otolaryngologists to 61.4 for obstetrician/gynecologists, with most other specialties ranging about 10 percent (Starfield, 1979).

Primary care practices also are widely believed to involve more patients who are continuing in care than those coming into care for the first time. Although several studies indicate that this is not always the case, it does hold for family medicine. In one study in the United States, family medicine was the only primary care specialty that consistently ranked first regarding the proportion of patients continuing in care to those coming for the first visit, in the proportion of patients without a specific referral from another practitioner, and in the proportion of patients referred *from* the practice to those seen on referral *into* the practice. Other specialties such as cardiovascular diseases, otolaryngology, and obstetrics–gynecology also ranked high on these attributes (Puskin, 1977).

Another characteristic that should distinguish primary care from specialty practitioners is the distribution of patient visits by familiarity with both the patient and the patient's problems. Both primary care physicians and specialists are expected to see new patients and "old" patients with old problems. Primary care practitioners should see more old patients with new problems, since they are responsible for the patient's care over time, regardless of the particular problem. Table 7.8 shows that the practices of certain specialists and primary care physicians differ in some re-

TABLE 7.8.   Distribution of Office Visits by Familiarity with Patient or Patient's Problems by Specialty of Physician: United States, 1985

	Percentage of Office Visits		
Specialty	New Patient	Old Patient New Problem	Old Patient Old Problem
Family/general practice	14.1	32.6	53.8
Internal medicine	15.3	22.9	61.8
Pediatrics	12.8	40.2	46.9
Medical subspecialists	16.9	10.1	73.0
General surgery	21.4	17.9	60.7
Obstetrics/gynecology	14.2	21.3	64.5
Other surgical specialties	25.1	8.1	66.8
Psychiatry	7.8	1.0	91.2
All other specialties	31.0	14.2	54.8

Source: NCHS, 1988.

TABLE 7.9.   Percentage of Visits in Which Patient Was Referred by Another Physician

		Children	*Adults*			
	All Ages	*under Age 15*	*15–24*	*25–44*	*45–64*	*65+*
Physician specialty						
All physicians	5.6	4.4	5.4	6.5	5.9	5.4
Primary care physicians	2.2	2.1	2.6	2.8	2.0	1.8
General/family practice	1.6	1.6	2.3	1.8	1.2	1.1
Internal medicine	4.1	14.2	a	5.5	3.7	2.7
Pediatrics	2.0	1.8	a	a	a	a
Other specialties	7.8	15.1	7.9	9.2	9.5	9.2

Source: National Ambulatory Medical Care Survey, 1985 (courtesy of James DeLozier).

aNumbers are too small to provide stable estimates.

spects. Surgical subspecialists, psychiatrists, and medical subspecialists see relatively fewer old patients with new problems than do other types of physicians. However, general internists (internal medicine) see fewer such patients than either general/family physicians or pediatricians, approximately the same percentage of such patients as obstetricians/gynecologists see and only slightly more than general surgeons (NCHS, 1988).

Thus, with the possible exception of the variety of problems presented by patients, primary care and specialty care in the United States cannot be differentiated by descriptive characteristics.

Tables 7.9 and 7.10 show the referral status of patients visiting primary care and nonprimary care physicians. Table 7.9 presents the percentage of specific visits made on referral from another physician. No more than 15 percent of these visits are made on such a referral, even for nonprimary care physicians as a group. A very small proportion (about 2 percent) of children's visits to both general/family physicians and pediatricians are made on referral. Visits to internists are two to three times more likely to be by referral than visits to general or family physicians, but even here the percentage does not exceed 5 percent except for the relatively few children seen by internists. Visits to specialists are two to four times as likely to be by referral than visits to primary care physicians. In the case of children, the percentage is even higher—15 percent of all visits.

TABLE 7.10.   Percentage of Visits in Which Patient Is Referred
to Another Physician, by Specialty of Physician: United States, 1985

Specialty	*All Ages*	*Children*	*Adults*
Total	3.2	2.2	3.4
Primary care	4.0	2.4	4.6
General/family practice	3.8	2.9	4.5
Internal medicine	5.3	—	5.0
Pediatrics	2.4	2.1	—
Other specialties	2.2	1.7	2.2

Source: National Ambulatory Medical Care Survey, 1985 (courtesy of James DeLozier).

Among nonprimary care specialists, the percentage of visits resulting from a specific referral from another physician varies considerably. The percentage is consistently low for visits to psychiatrists and obstetrician/gynecologists and consistently high for visits to urologists and neurologic surgeons. About one-third of visits to cardiovascular specialists by young adults are by referral, but the proportion falls progressively with age to 3.6 percent among those over age sixty-five. This probably reflects the greater frequency of cardiovascular disease in older people and therefore greater familiarity of primary care physicians with those problems as well as greater confidence in dealing with them. The proportion of visits to other subspecialists that derive from a referral from other physicians is generally between 10 and 20 percent and declines with age of patient for most subspecialists. This could be a result either of more self-referrals of older patients to subspecialists or of subspecialists retaining them within their practices, or both.

The NAMCS study yields information relating to whether or not a specific visit occurred because of a referral from another physician. If a patient had been referred earlier and returned to the specialist for continuing care, the visit was not counted as a referral. A national study of over twenty specialties in the mid-1970s used approximately the same techniques as the NAMCS, but the specific questions were different. The study asked how the physician initially obtained the patient for the problem that prompted the specific visit, and further inquired whether the patient had been referred just to obtain an opinion or advice or was referred permanently. For physicians in all specialties a larger proportion of patients were referred permanently than were sent for advice or an opinion, and as expected, the percentages of patients seen on referral as a whole was considerably greater than that found in the NAMCS survey. However, findings resembled those of the NAMCS in showing that internists referred a greater percentage of patients than did general or family physicians. The specialists with the highest percentage of in-referrals were neurosurgeons, and the specialists with the lowest percentage of in-referrals were obstetricians/gynecologists, ophthalmologists, and dermatologists, largely because of a much smaller percentage of patients referred permanently (Robert Wood Johnson Foundation, 1982).

Table 7.10 presents the percentage of visits that result in a referral to another physician. This percentage is smaller for children than for adults and generally similar for general or family physicians and pediatricians (in visits by children) and for general or family physicians and internists (in visits by adults). Although the differences in referral rates among primary care physicians are neither striking nor statistically significant, generalists refer fewer young adults than do internists; as the age increases, the difference narrows until age sixty-five, when generalists refer slightly more patients than do internists. Nonprimary care specialists refer a smaller percentage of both their adult and child patients than primary care physicians, but the generally low rates of referral by all types of physicians is striking.

In the study conducted in the mid-1970s, the frequency of referrals out of the practice was considerably greater than that in the NAMCS, but similar in that rates of referral by primary care physicians (about 7 percent in family practice, 8 percent in pediatrics, and 16 percent in internal medicine) were much lower than those for other physicians. Rates of referral by other types of physicians ranged from about 5 percent for ophthalmologists to 23 percent for neurologists. For most nonprimary care physicians, the percentage of encounters referred permanently was three to five times greater than the percentage referred for consultation only. For primary care physicians, the percentage referred permanently (4 to 10 percent) was about twice the percentage referred for a consultation only (2 to 6 percent) (Robert Wood Johnson Foundation, 1982).

The National Ambulatory Care Survey also obtains information on the proportion of visits in which the patient is referred back to the referring physician. For primary care physicians, these rates are under 1 percent for both adults and children. The highest proportion of visits resulting in a referral to return to the referring physician is from visits to neurologic surgeons (9.3 percent) and cardiovascular specialists (5.4 percent). For all specialists, however, the percentage of visits resulting in a return to a referring physician is smaller than that of patients referred from another physician, which suggests that, overall, a substantial proportion of patients who are referred to another physician do not return to the referring physician, at least for the problem prompting the visit.

These data on the content of care in practice in the United States indicate that office-based primary care and nonprimary care physicians do not differ very much with regard to several important clinical characteristics of visits, such as the type of reason for visit and referral characteristics. It is possible that inclusion of hospital-based specialists would reveal greater differences, but no data provide information for such a comparison.

Challenges for Primary Care in the United States

The information presented at the beginning of this chapter reveals an apparent blurring of the roles of primary care and nonprimary care physicians in the United States. The proportion of "true" primary care physicians, which includes family and general practitioners, general pediatricians and nonsubspecializing internists, is declining while the proportion of subspecialists is increasing. Moreover, the practices of primary care physicians and office-based nonprimary care physicians are not very different, at least concerning the stage of differentiation of patients' problems and their referral characteristics.

The historical trend of specialization that started in the early decades of the century was fueled by the large expansion of biomedical research capacity in the late 1940s and early 1950s. The availability of research grants led directly to the growth of biomedically oriented faculty in medical schools, with consequent shifts in curriculum provided to medical students and to changes in the type of role models to which they were exposed during training (Lewis and Sheps, 1983; Starr, 1982). Despite widespread (although not unanimous) agreement concerning the need for more primary care orientation in medical practice, medical school graduates increasingly choose specialty careers rather than primary care careers (IOM, 1989, p. 27), and the proportion of physicians who focus on primary care practice is still falling (IOM, 1989). Medical students may be choosing subspecialty careers rather than primary care because they perceive that primary care would be too demanding of their time and skills. In particular, the challenges of availability and commitment to patients over time may be responsibilities that interfere with the pursuit of the physician's own desired life-style (Schwartz, 1989). The problem is exacerbated by the declining productivity of physicians, especially in the case of primary care physicians (Freiman and Marder, 1984). The distribution of physicians continues to be unrelated to the needs of the population. Rural areas and areas of cities with large concentrations of socioeconomically deprived people face larger deficits than other areas of the country in primary care, especially that provided by family and general practitioners (Barnett and Midtling, 1989).

In the late 1970s, the GMENAC (*Graduate Medical Education National Advisory Committee*) (USDHHS, 1980) estimated the future of supply of specialists and primary care physicians, contrasting it with the projected demand for their services. It concluded that the overall supply, including that of almost all specialists, would be in excess of the demand, but it estimated that the supply and demand for primary care physicians would be in balance. Since then, the demand for specialty services has increased more than the supply of specialists (Schwartz et al., 1988). In contrast, new developments indicate that more primary care physicians will be needed than had been projected. New conditions such as acquired immune deficiency syndrome (AIDS) are placing increased demands on primary care physicians. As the baby boom generation ages, the burden of increased morbidity of the elderly will place even more demand on primary care physicians. The increase in managed care (see the chapter on organization and financing of services in the United States), in which primary care physicians serve as gatekeepers, adds to the increased demand for them. The GMENAC estimated that nonphysician health personnel would provide a portion of primary care. However, the projected growth of and demand for this type of practitioners has failed to materialize.

The declining supply of primary care physicians is eroding the provision of primary care in the United States. Despite evidence that specialty-trained physicians are providing "primary care" (Aiken et al., 1979), specialists do not perform as well as primary care physicians in achieving the four key characteristics of primary care. The hospital is an inappropriate place for training in primary care. Except in family medicine, residency training is overwhelmingly in the inpatient setting, where patients suffer from serious acute conditions or acute exacerbations of chronic conditions, mostly unconnected with primary care. There is little "bloc" in which to learn the challenges of the care of patients over time, and little opportunity to learn what to do with the symptoms that comprise the bulk of primary care practice. There is also little exposure to chronic, non–life-threatening ailments, be they physical or psychosocial. Residents in training have little opportunity to learn the challenges of preventive care, and no opportunity to learn how to deliver care in which they are both the gatekeepers and the integrators of the patient's health care experiences over time. Although these problems are greater in specialty training than in primary care training, the realities of hospital-based training in primary care lead to the same types of deficiencies as are present in the subspecialties. For example, only a small minority of residencies in internal medicine include experiences in outpatient settings other than hospital clinics, and the amount of time spent in the outpatient setting is generally less than 20 percent of the training period (Clare et al., 1987; Andersen et al., 1990).

Financial aspects of the system of educating physicians militate against changing the focus on specialty training. The financing of medical education depends heavily on income from the provision of clinical services, and technologically oriented subspecialties generate more clinical income than the primary care specialties do. More than 80 percent of the costs of residency training come from the provision of clinical care; income from clinical services grew from 12.2 percent of the revenues of medical schools in 1970 to 37.6 percent in 1987. On average, only 46 percent of the activities of internal medicine faculties and only 31 percent for the faculties in family practice are supported by clinically derived income (Friedman, 1986). Because primary care residencies generate less revenue than other specialties, teaching hospitals have found it increasingly difficult to support them (Barnett and Midtling, 1989). Funding targeted specifically at the support of primary care residencies has not kept pace with inflation, and is threatened with discontinuation because of the perceived excess of physicians and governmental budget deficits (Barnett and Midtling, 1989).

To narrow the gap in support of primary care and specialty programs, proposals have been made to increase the reimbursement for services provided by primary care physicians and to decrease those for specialists (Hsiao et al., 1988). However, the changes will be too small to reverse the incentives for specialty practice, which will change only with major alterations in the financing of graduate medical education (Barnett and Midtling, 1989; IOM, 1989). It may also be useful to consider methods of payment that intentionally reward the provision of primary care by primary care physicians. For example, primary care physicians might be paid fee-for-service while nonprimary care physicians would be paid a salary (Menken, 1988). It will also be necessary to reduce inappropriate visits to subspecialists by requiring patients to seek the care of primary care physicians before they consult specialists, as an increasing number of health insurance plans and health services organizations do.

Specialists cannot provide adequate primary care. Their hospital-based training does not expose them either to the challenges of primary care or to its principles. In addition, when nonprimary care specialists provide primary care it jeopardizes competence in their own specialty. Maintenance of expertise requires a critical volume of care in the specialty; dilution of activities with care unrelated to their special expertise reduces the amount of care concentrated in the specialty. Competition from an increasing number of similar specialists reduces referrals from primary care and hence further decreases the critical volume of specialty problems in their practices.

The increasing demand for specialty services in the United States is consistent with a historical emphasis on freedom of choice of physician and a public belief that specialized care is better care. As is shown in Chapter 12, quality of care is heavily determined not only by the volume of practice in the specialty but also by the extent of training in the particular subject being examined and the nature of the organization in which the physician works. As long as quality of care is measured by the adequacy of management for specific diseases that are largely in the purview of subspecialists, primary care physicians will be found deficient. However, as this chapter has shown, the focus of primary care is not on diagnoses but rather on assessing and managing symptoms. When patients with symptoms seek care directly from specialists. they make implicit assumptions that the symptoms are associated with a disease in the purview of the specialist. If their judgment is correct, they are more likely to receive superior care than they would have from a primary care physician, provided the problem is seen infrequently in primary care. If they are incorrect, the care received will be inappropriate. In both cases, it will be more costly, as is discussed in Chapter 9.

The data from NAMCS indicate many similarities between primary care and nonprimary care specialists. Although specialists are more likely than primary care physicians to receive patients on referral and less likely to refer patients to other physicians, the differences are not marked. Indeed, specialty care and primary care are blurred in the United States. At best, this situation contributes to the unusually high costs of the U.S. health system, at least compared with countries of similar industrialization and culture heritage. At worst, it also may be compromising quality of care (Schroeder, 1985; Menken, 1988).

References

Aiken L, Lewis C, Craig J, Mendenhall R, Blendon R, and Rogers D. The contribution of specialists to the delivery of primary care: a new perspective. N Engl J Med 1979; 300:1363–70.

American Medical Association. *Physician Characteristics Distribution in the U.S.* Chicago. 1992.

American Medical Association. *Physician Characteristics Distribution in the U.S.* Chicago. 1984.

Andersen R, Lyttle C, Kohrman C, Levey G, Neymarc K, and Schmidt C. National study of internal medicine manpower: XVII. Changes in the characteristics of internal medicine residents and their training programs, 1988–1989. Ann Int Med 1990; 113:243–49.

Barnett P and Midtling J. Public policy and the supply of primary care physicians. JAMA 1989; 262:2864–88.

Bureau of Health Professions. *Projections of Physician Supply in the U.S.* HRP-0906330 ODAM Report No. 3-85, Hyattsville, Md March 1985.

Clare FL, Spratley E, Schwab P, and Iglehart J. Data watch: trends in health personnel. Health Affairs 1987; 6:90–103.

Freiman M and Marder W. Changes in hours worked by physicians, 1970–80. Am J Public Health. 1984; 74:1348–52.

Friedman R. Family practice and general internal medicine: what kind of cooperation makes sense? JAMA 1986; 255:2644–46.

Glandon G and Shapiro R. (eds.) *Profile of Medical Practice, 1980.* American Medical Association, Chicago, 1980.

Hsiao W, Braun P, Yntema D, and Becker E. Estimating physicians' work for a resource-based relative-value scale. N Engl J Med 1988; 319:835–41.

Institute of Medicine. Primary Care Physicians: Financing Their Graduate Medical Education in Ambulatory Settings. Washington D.C., National Academy Press, 1989.

Knowles J. The quantity and quality of medical manpower: a review of medicine's current efforts. J Med Educ 44:81–118, 1969.

Lewis IJ and Sheps CG. The Sick Citadel: The American Academic Medical Center and the Public Interest. Cambridge, Mass., Oelgeschlager, Gunn, and Hain, 1983.

McCrindle B, DeAngelis C, and Starfield B. Subspecialization within pediatrics practice: a broader spectrum. Pediatric Research 1989; 25:135A.

Menken M. Generalism and specialism revisited: the case of neurology. Health Affairs 1988; 7:115–23.

NCHS (National Center for Health Statistics). Health United States, 1989, Hyattsville, Md., Public Health Service, 1990, Table 64.

NCHS (National Center for Health Statistics). L Lawrence and T McLemore. 1981 Summary: National Ambulatory Medical Care Survey. Advance Data No 88. DHHS Publication No. (PHS) 83-1250. Hyattsville, Md., March 16, 1983.

NCHS (National Center for Health Statistics). C Nelson and T McLemore. The National Ambulatory Medical Survey: United States 1975–81 and 1988 Trends. Vital and Health Statistics. Series 13 No. 93. PHHS Pub. No. (88-1754). Public Health Service. Washington, D.C., U.S. Government Printing Office, 1988.

Puskin D. Patterns of Ambulatory Medical Care in the United States: An Analysis of the National Ambulatory Medical Care Survey. Doctoral thesis, The Johns Hopkins University School of Hygiene and Public Health, Division of Health Care Organization, Baltimore, 1977, pp. 68–74.

Robert Wood Johnson Foundation. Medical Practice in the United States. Princeton, N.J., 1982.

Rousselot L. Federal efforts to influence physician education, specialization distribution projections and options. Am J Med 1973, 55:123–30.

Rowley B, Baldwin D, McGuire M, Etzel S, and O'Leary C. Graduate medical education in the United States. JAMA 1990; 264:822–32.

Schroeder S. The making of a medical generalist. Health Affairs 1985; 4:22–46.

Schwartz R, Jarecky R, Strodel W, Haley J, Young B, and Griffen W. Controllable lifestyle: a new factor in career choice by medical students. Academic Medicine 1989: 64:606–9.

Schwartz W, Williams A, Newhouse J, and Witsberger C. Are we training too many medical subspecialists? JAMA 1988; 259:233–39.

Starfield B: Measuring the attainment of primary care. J Med Educ 1979; 54:361–69.

Starr P. The Social Transformation of American Medicine. New York, Basic Books, 1982, pp. 353–58.

Stevens R. Graduate medical education: a continuing history. J Med Educ 1978; 53:1–18.

U.S. Department of Health, Education and Welfare. *A Report to the President and Congress on the Status of Health Professions Personnel in the United States.* DHEW Publication No. (HRA) 78-93, Washington, D.C. 1978.

USDHHS. Graduate Medical Education Advisory Committee to the Secretary. GMENAC Summary Report. 1. Pub. No. (HRA) 81-651. Washington, D.C., Health Resources Administration. 1980.

Wechsler H. Handbook of Medical Specialties. New York. Human Sciences Press, 1976.

8

Organization, Financing, and Access to Services

The United States is almost unique among industrialized nations in lacking a national health system to guide the organization, delivery, and financing of health services in the nation. The system of organization and delivery is largely left to the private sector, with government assuming responsibility for financing of care for certain segments of the population.

Physicians generally are self-employed, and most hospitals are owned by private corporations, some of which operate their hospitals to provide a profit for their owners. However, a substantial amount of care provided in the private sector is paid for by public funds. Various levels of government provide some care directly, such as for the military and its dependents, for many of the very poor, and for many American Indians.

The major distinguishing characteristics of the predominant forms of health services in the United States concern three particular structural features of the health services system: the type of organization, the method of reimbursement, and the size and nature of the population for which it assumes responsibility. This chapter reviews these characteristics, indicates how they may be changing over time, and considers their impact on access to services.

Although most of the civilian population of the United States receives medical care from a physician who works alone or with a partner, group practices began developing as long ago as the 1880s. In the 1920s the rate of growth began to increase (Starr, 1982). The standard definition of a group practice is "the application of medical services by three or more physicians formally organized to provide medical care, consultation, diagnosis, and/or treatment through the joint use of equipment, records, and personnel, and with income from medical practice distributed according to some prearranged plan" (Havlicek, 1990). By the late 1980s, almost one-third of physicians worked in group practices. However, fewer than a third of all primary care physicians were in groups: 24 percent of internists, 26 percent of family of general practitioners, and 31 percent of pediatricians (Havlicek, 1990).

Multispecialty groups are, on average, much larger than single specialty groups. The most common single specialty groups are composed of radiologists, internists, obstetricians–gynecologists, or pediatricians.

The organizational arrangement in which physicians work is not necessarily linked to the method by which they are paid. Physicians can be paid a fee-for-service, a salary, or capitation.

Fee-for-Service: This is the most common method. Practitioners are reimbursed for each service or procedure provided to patients. The more the services and the greater the number and complexity of services, the higher the payment. Sometimes patients must pay the physician at the time of service or in response to subsequent billing. The patient may be reimbursed by insurance companies (indemnity plans). In other arrangements, physicians may be paid directly by insurance companies or by governmental agencies.

Salary: In this method of payment physicians receive a fixed sum of money based on the amount of time they devote to their professional work.

Capitation: In this method of payment, practitioners are paid according to the number of patients assigned to receive services from them. With this form of remuneration, patients must be enrolled for a defined period of time, which may vary from a month to a year. Practitioners can receive no more than a set sum of money for providing services unless the insuring or managing agents provide for added payments under predetermined circumstances or for services they wish to encourage. Capitation is the method of payment that is most conducive to achieving longitudinality in primary care, since payments are based on enrollment of a defined panel of patients over a period of time. Although primary care physicians are more likely to be paid by capitation than specialists, only a small percentage of primary care physicians are paid by this mechanism.

This conventional structure is becoming more complex as a result of a variety of types of organizations of physicians, fiscal intermediaries, and management contractors and increasingly complex mechanisms of paying for services. Physicians are now working for a variety of types of organizations and being paid by a variety of mechanisms.

Insurance companies, employers who act as their own insurers, and government agencies are increasingly contracting with management companies that control expenditures by placing a variety of constraints on physicians. Sometimes there is more than one of these "middle tiers," which may consist of physician groups, a hospital medical staff, or an entity formed specifically to provide management services. Typically, the middle tier imposes penalties or offers incentives to the practitioners to keep expenditures within limits imposed by the insurance available to pay for services and sometimes provide a profit for the insurer or management organization.

The next two sections of the chapter, respectively, describe the major new forms of health services organization and ways of financing health services.

New Forms of Health Services Organizations

Organizational arrangements that are assuming increasing importance in the United States are characterized by various combinations and permutations of organizational format and reimbursement. The conventional typology divides these combinations into three major types: health maintenance organizations (HMOs), preferred provider organizations (PPOs), and managed indemnity plans (MIPs). This section describes these forms of care.

Health Maintenance Organizations (HMOs)

In the past forty years, prepayment for services has been gradually increasing in frequency. In this form of reimbursement, third party payers contract with physicians or physician groups. In return for prepayment, the providers agree to furnish a package of services defined in a contract. In some forms of prepayment, these physicians are also at financial risk of costs of care that result from referrals or hospitalizations, although most arrangements exempt certain "catastrophic" costs over which the physician has no control. These arrangements are known as health maintenance organizations (HMOs).

In 1932 the Commission on the Costs of Medical Care recommended that health care be provided by organized groups of health professionals, preferably in a hospital setting, on a prepayment basis. Although care provided by such organized groups is a feature of many foreign health care systems, only in the United States are they organized within the private sector and reimbursed largely through private insurance plans.

HMOs are identified by a prepayment for care. Payment is fixed in advance and depends on the number (and sometimes type) of enrolled patients. Physicians who work in the HMO must control health care expenditures; if more care is provided than the HMO anticipates, the HMO incurs a deficit. On the other hand, if too little care is provided, the HMO may be suspected of not providing necessary care.

Originally, HMOs just provided clinical services. In current terminology, however, the HMO is often the fiscal agent that contracts either with a management tier or directly with a medical care group to provide services for its defined population. The physicians or physician groups within an HMO are of several types:

- *Staff Model:* The physicians work directly for the HMO, on salary but sometimes with a bonus depending on the HMO's earnings.
- *Group Model:* The HMO contracts with a separate physician group to provide its services. These are also known as prepaid group practices (PPGPs), and physicians may be paid a salary or a capitation. In network model HMOs, the contract is with multiple physicians or group practices instead of just one.
- *Individual Practice Association (IPA):* The HMO contracts with individual physicians in independent practice or with a collective of physicians whose members work independently, usually in solo practice or single specialty groups. IPA plans typically require primary care physicians to act as gatekeepers for their patients. This role requires them to approve referrals, admissions, high-cost procedures, and tests. A capitation fee is paid to the HMO for each person enrolled. Physicians may be paid a capitation for primary care services or may be reimbursed according to a fixed fee schedule for services rendered; the fees are typically based on a percentage of the physician's usual fee. In a network model IPA, primary care physicians may receive additional capitation for referral and ancillary services arranged by the primary care

physician and paid for by fee-for-service or by a prearranged payment schedule. Usually a portion of the fee or the capitation payment is withheld by the IPA. If utilization and costs are in line with expectations, the withheld amount is returned to the physicians.

In both the group model and IPA, the physicians are usually free to see patients outside of the HMO in addition to those enrolled in the HMO.

In a *point-of-service* plan, patients are free to seek care from non-HMO physicians, but pay considerably more for doing so.

Another type of HMO, the social HMO or S/HMO, has been developed and tested by the federal government as a means of integrating the long-term care and acute care services to the chronically ill elderly.

Managed Indemnity Plans (MIPs) and Preferred Provider Organizations (PPOs)

A managed indemnity plan is a conventional (indemnity) fee-for-service arrangement in which the use of services and procedures is carefully monitored.

A PPO is an administrative entity that contracts with employers or insurers to provide services to individuals for a negotiated fee, which is usually discounted below the prevailing level. In a PPO, patients may choose the physician, regardless of whether or not the physician participates in the PPO. If they use a non-PPO physician, they must pay part of the fee (coinsurance or a deductible) out-of-pocket. In an exclusive provider organization (EPO) use of non-PPO physicians is not covered except in emergency situations. The EPO, therefore, is similar to the IPA, except that EPO pays its physicians a negotiated fee-for-service whereas the IPA usually shares risk with its physicians (and always with its primary care physicians) by withholding a portion of the capitation or fee-for-service until all expenditures are tallied. This portion is not returned unless financial targets are met.

Integral to PPOs is a managing organization that provides certain administrative functions to monitor the use of services. These may include preadmission certification (i.e., approval of elective admissions), second opinions before surgery, certification of treatment plans for certain nonemergency services such as mental health services, and review of the medical care that is provided.

Free-Standing EmergiCenters (FECs)

FECs are free-standing walk-in facilities for the treatment of nonemergency medical conditions. Most are equipped to provide only minor and non–life-threatening care and some routine health services. They do not require an appointment and are open during times when most physicians' offices and hospital clinics are closed. Most require immediate payment by cash or credit card. Between 1978 and 1984, the average annual growth of FECs was 71 percent; by the end of 1983, estimates of the number nationwide ranged from 900 to 1,300. However, since 1983, the growth in the number of FECs and patient volume has abated; many FECs are not profitable and are restructuring or closing. Originally about three of every four FECs were

TABLE 8.1.   Types of Health Services Organizations, by Method of Payment to Physicians and Source of Payment

| Source of Payment | Prenegotiated Fee | | | |
| | No Enrollment[a] | | | |
	Salary	Fee-for-Service[b]	Enrollment[c]	Reimbursed Fee
Out-of-pocket			Individual contracts for specified types of care (e.g., obstetric, dental)	Traditional ambulatory services Free-standing emergi-centers
Private insurance	Corporate health centers	PPOs MIPs	HMOs (PPGPs and IPAs)	Most inpatient services Certain ambulatory services
Tax revenues	Some governmental health centers, community health centers	PPOs	Prepaid Medicaid Programs	Medicaid Champus
Federal insurance	Veterans Administration facilities	PPOs (through Medicare)	HMOs (special arrangements)	Medicare

[a]For purposes of this table, enrollment means that individual names are provided to physicians, which, in the aggregate, constitutes a practice roster.
[b]Individual physicians paid fee for service, although subject to utilization review.
[c]Payments to physician may be by capitation or salary (as in PPGPs and some IPAs) or by fee-for-service.

owned by physicians, largely for profit. Currently they are being taken over by large national firms involved in other types of organizations for managing care.

Table 8.1 summarizes the different ways in which medical services are organized and financed. Table 8.2 summarizes the results of many studies on the relationship between type of organization and various aspects of access to and quality of care. As Table 8.2 notes, those characteristics especially related to primary care services—namely, preventive visits and visits for illnesses common in primary care—are best achieved in HMO-type organizations (Stoline and Weiner, 1988).

Trends in the Development of New Organizational Formats

During the 1930s and especially the 1940s, major societal changes made access to medical care a societal priority. Labor–management negotiations generally included bargaining over health insurance as part of employee benefits; an increasing proportion of the benefit package consisted of employee and family coverage for health care costs. Moreover, the federal government assumed an increasing responsibility for paying for care of unemployed individuals unable to afford it. During this period, the precursors of PPOs were developed, with the first prototype in 1934. The forerunners of the IPA were also initiated during this period, generally by county medical societies in the form of foundations for medical care. Kaiser Industries, faced with the need for medical services to care for the influx of workers in

TABLE 8.2. Relationship Between Type of Health Insurance Plan and Characteristics Related to Primary Care and Specialty Care

Characteristic of Medical Care	Type of Health Insurance Plan				
	Conventional Fee-for-Service Insurance Plan	Managed Indemnity Plan	Preferred Provider Organization	IPA/Network Model HMO	Staff/Group Model HMO
More highly related to primary care					
Preventive visits	– –	– –	+	+ +	+ +
Illness-related visits to primary care physician	–	–	+	+ +	+ +
More highly related to specialty care					
Visits to specialists	+ +	+	+	–	– –
Use of diagnostic tests	+ +	–	–	–	– –
Rate of surgery	+ +	–	–	–	– –
Admission to hospital	+ +	–	–	–	– –

Source: Stoline and Weiner, 1988.

Key: Expected impact of plan on consumer use/physician practice (all else equal): – – should decrease, – tends to decrease, ? direction of effect not clear, + tends to increase, + + should increase.

113

defense industries during World War II, developed the first large-scale prepaid group practice HMO.

These "alternative" delivery systems grew very slowly; in the 1970s, fewer than 5 percent of the population and physicians were involved in them. Medical costs rapidly increased, fueled by a growth in technology in the form of diagnostic and therapeutic tests and by procedures and a system of reimbursement (fee-for-service) that simply passed the costs on to patients and their insurers. This provided impetus for new thinking about prepayment and cost controls.

In 1982, the Tax Equity and Fiscal Responsibility Act (TEFRA) encouraged the enrollment of the elderly in HMOs. Between 1981 and 1986, the number of HMOs increased by 145 percent and the number of individuals enrolled increased by 130 percent, that is, from 10.3 to 23.7 million people or to about 10 percent of the population (Owens, 1987). From 1984 to mid-1987 enrollment grew at an annual rate of more than 20 percent. The IPA type of HMO experienced the largest growth. During the period from 1981 to 1985, aggregate HMO enrollment doubled, while IPA enrollment quadrupled (Dalton, 1987). At the end of 1985, 51 percent of the HMOs were IPAs; 19 percent were network HMOs, 17 percent were group HMOs, and 13 percent were staff HMOs. By late 1987, however, there was evidence of a marked slowing in the rate of increase in enrollments in HMOs. Many corporations eliminated health insurance coverage for care in HMOs. At least sixteen HMOs disappeared in 1987 because of mergers or business failures. The HMO type of care is increasingly becoming dominated by a few large national organizations, some being the original nonprofit predecessors, such as the Kaiser Health Plan and the Health Insurance Plan of New York, and others the newer profit-seeking plans, such as United Healthcare and US Health Care. By 1989, HMO enrollment had reached only 32.5 million people and there were fewer HMO organizations than in 1987 (Gruber et al., 1989).

Before 1983, only 20 PPOs were operational in the United States, whereas by 1986 there were 413 in 41 states. From June 1985 to June 1986 alone there was an increase of 59 percent in the number of PPOs (Dalton, 1987). The total number of individuals served by PPOs is unknown, but various estimates have been made. By mid-1987, the American Association of Preferred Provider Organizations (AAPPO) reported 674 PPOs, involving 32 million people with PPO options. Although hospitals and Blue Cross/Blue Shield Plans had the largest market share in 1986, commercial insurers and investor-owned PPOs were growing at the most rapid rate (deLissovoy et al., 1987). Three states—California, Colorado, Florida—accounted for 65 percent of PPO enrollees. As more states strike down statutes prohibiting selective contracting, PPOs are likely to spread to other parts of the country (deLissovoy et al., 1987).

By 1987, roughly one-fourth of the U.S. population had their health care financed and delivered through an HMO or PPO. This marked growth in alternative delivery formats has been accompanied by notable shifts in the ownership and management of medical care organizations. Initially, HMOs were established as not-for-profit organizations. The federal Health Maintenance Organization Act of 1973, which was implemented in 1976, gave impetus for the development of HMOs by providing loans and eliminated the federal loan program of 1973, thus encouraging HMOs to seek private-sector investment capital for start-up cost. Most of the applications were from not-for-profit organizations. In the 1980s, the national administration encouraged a movement toward marketplace competition; this resulted in the entry into the health field of entrepreneurs and venture capitalists. The rapid growth of for-profit national health care corporations also spurred the entry into the field of insurance companies, who had to protect their own indemnity plans from the new competition. From 1981 to the end of 1985, the percentage of HMOs that were for-profit had grown from 18 to 51 percent. Although the majority (54 percent as of 1986) of PPOs are operated by hospitals and physician groups, it is likely that insurance companies will capture an increasing share of the market so that these organizations, too, will be dominated by a profit motive (Dalton, 1987).

The hybridization of health services organizations, even within the category of HMOs, is becoming so marked that some analysts have proposed new typologies to characterize them (Welch et al., 1990). Three major structural characteristics are key to defining five types of organizations:

- The way in which physicians are paid for primary care services (a *financing* characteristic).
- The nature of the risk or reward to primary care physicians above their basic payment method (a *financing* characteristic).
- Whether only HMO patients are seen or the organization sees fee-for-service patients covered by traditional indemnity insurance only or in combination with HMO patients (a structural characteristic concerning *definition of the eligible population*).
- The size and nature of the risk pool used to share the risk or reward (a structural characteristic concerning *definition of the eligible population*).
- The nature of the organization's contract with its management tier, if any (an *organizational* characteristic).

Each of these characteristics has *at least* two possible forms; the large number of combinations indicates the large number of types of organizations that are likely to exist within the country.

In mid-1988, Welch and colleagues surveyed all HMOs in existence at that time and classified them in two different ways. First, the HMOs were characterized by the nature of their financial incentives and their risk pool. Second, they were characterized by the number of management tiers. Table 8.3 compares the distribution of the HMOs categorized by the more conventional method with the newer categorization. As the table indicates, the new typology shows greater diversity among organizational structures than does the more conventional typology. Table 8.4 provides information on the performance of these different types, as assessed by the extent of hospitalization of their patients, the average number of specialty visits per enrollee, the profitability of the organization, and the rate of growth in enrollment. In the newer characterization, the range of differences in number of specialty visits is greater than in the older typology, reflecting the great impact of financial incentives and management controls on the likelihood of referrals by primary care physicians and the ability of patients to seek unrestricted care directly from specialists.

Other studies have also demonstrated that placing primary care physicians at

TABLE 8.3. New Typologies Compared with More Conventional Typologies
(percent of HMO enrollment)

| New Typology | Conventional Typology (Interstudy) | | | | |
	Staff	Group	Network	IPA	All HMOs
Incentive based					
Prepaid group practice	12.8	21.3	1.5	0.0	35.7
Salary IPA	0.2	3.5	0.6	0.1	4.4
Capitation IPA	0.5	0.3	4.8	13.9	19.4
FFS IPA with subgroup	0.0	0.0	3.2	11.1	14.4
Foundation-type IPA	0.0	0.3	0.0	13.4	13.7
Missing	0.0	0.2	7.9	4.3	12.3
Total	13.5	25.7	18.0	42.8	100.0
Organization structure					
Prepaid group practice	12.8	21.3	1.5	0.0	35.7
Two-tiered IPA with subgroups risk pools	0.0	0.0	0.7	17.7	18.4
Two-tiered IPA with a single risk pool	0.1	0.1	0.7	12.8	13.6
Three-tiered IPA with subgroups risk pools	0.5	0.5	13.7	5.6	20.2
Three-tiered IPA with a single risk pool	0.0	2.9	0.2	2.6	5.7
Missing	0.0	1.0	1.2	4.2	6.4
Total	13.5	25.7	18.0	42.8	100.0

Source: Welch et al., 1990.

financial risk for costs of services alters the frequency of use of services by patients. For example, when physicians are paid a capitation and are placed in risk pools, their HMO patients receive fewer discretionary tests but the same number of indicated preventive tests as do those seen under fee-for-service arrangements (Clancy and Hillner, 1989). Other investigators found that placing physicians in risk pools reduces the frequency of visits by patients (Hillman et al., 1989).

Thus, changes in the very important structural characteristics concerning organization and financing of services have major implications concerning use of services by patients. When physicians' incomes are directly influenced by the number and types of services they give their patients, there is great potential for underuse of services, even well-indicated ones. With no incentives, the potential for overuse and possible harm to patients' health from unnecessary tests, procedures, and therapies is increased.

The trade-off between these alternatives is likely to be most acute in individuals with the greatest health needs. Health care organizations competing for enrolled populations while attempting to maximize profits cannot be expected to welcome individuals likely to require a relatively large number of services. Since nonprofit organizations must compete for their survival with profit-making organizations, they are no more likely to welcome high-risk populations as enrollees than are for-profit organizations. Experience with a variety of programs to encourage enrollment of employers in health care plans clearly shows that programs generally adopt techniques which will minimize the possibility of adverse risk selection. Such techniques include the exclusion of certain types of employment with relatively high

TABLE 8.4. Comparison of Typologies in Terms of Performance Measures

Typology	Hospital Days/1,000	Specialty Visits	Whether Profitable	Enrollment Growth (%)	Plan Age	N
Interstudy (Conventional)						
Staff	376	1.53	.48	3.9	16.2	27
Group	373	1.41	.94	7.7	35.6	29
Network	309	1.66	.53	16.6	10.3	39
IPA	359	1.82	.68	21.9	6.7	165
Incentive based						
Prepaid group practice	375	1.44	.75	5.8	31.0	41
Salary IPA	375	1.46	.77	18.2	8.2	18
Capitation IPA	359	1.75	.58	22.7	7.5	92
FFS IPA with subgroup	343	1.58	.86	29.3	7.1	34
Foundation-type IPA	369	2.01	.75	19.5	7.1	47
Organization structure						
Prepaid group practice	375	1.44	.75	5.8	31.0	41
Two-tiered IPA with sub-groups risk pools	357	1.51	.74	29.7	6.7	69
Two-tiered IPA with a single risk pool	373	1.79	.65	20.9	6.8	64
Three-tiered IPA with sub-groups risk pools	312	1.72	.62	12.2	10.0	40
Three-tiered IPA with a single risk pool	360	1.96	.84	23.2	7.5	22

Source: Welch et al., 1990.

risks, such as those in construction firms, and preenrollment health screening to detect existing medical conditions and rejection of those who have them. Since marginal and low-paid workers and their families are more likely to be excluded, their children will be more disadvantaged in the future than they have been in the recent past. Chronically ill individuals, poor risks for any prebudgeted health care organizations, might also be at increased disadvantage. In the past, public agencies have sought to overcome barriers to access to care for those in high-risk groups by passing legislation barring such stratagems. Whether or not state legislatures will develop new strategies, such as insurance pools supported by a tax on employers to pay for care of the uninsured, remains to be seen; such options are currently under consideration in several states.

Increasing hybridization may also compromise the superior performance of staff and group model HMOs. A variety of studies have shown that these HMOs provide higher-quality care, as measured by rates of performance of preventive procedures and lower hospitalization rates, with no adverse effects on health (Luft, 1981; Retchin and Brown, 1990). Between 1986 and 1988, the percentage of HMOs requiring a copayment for primary care visits increased from 38 to 48 percent (Gold and Hodges, 1989). Competitive pressures to reduce current costs by imposing financial barriers to services with benefits that will be realized only in the future could eliminate the historical advantage of enrollees in such organizations.

Financing of Health Services

Financing arrangements are often very complicated. For example, the managing tier may pay its practitioners a fee-for-service even though it may be paid a capitation fee by the fiscal agent. Or it may pay some of its practitioners a fee-for-service and others a capitation or salary. Further complications result when different types of providers are paid by different methods. For example, primary care physicians may be paid by capitation or salary while specialist services or inpatient care is paid for by the fee-for-service arrangement. Sometimes the primary care physician or groups of primary care physicians are placed at risk for expenditures generated by specialists or hospitals to whom they refer their patients, even though they have no direct control over the level of these expenditures. Thus, the gatekeeper function of primary care is a critical function in risk-sharing arrangements.

When there is only one payer and one provider who is paid a fee-for-service, there is no administrative reason to define a population. Any insurance plan, however, must define its "risk pool" in order to stay within the budget determined by the money available for services. For insurance companies, this is the premium it collects from those who purchase health insurance. The more complex the organizational structure, the more complex is this definition of the risk pool. Fiscal agents may contract with more than one management tier, and a management tier may contract with more than one provider or provider group. Each of these organizational levels may be placed at risk in the sense that they are given a fixed budget to provide services to the defined population. The smaller the defined population, the greater the risk incurred, since expenditures may exceed the prebudgeted amount if only a few individuals in the population need very costly services within the budget period. That is, there is not a large enough pool to "spread the risk." Therefore, there is considerable incentive to develop larger and larger services organizations, at least within any given management tier. The remainder of this section describes the major ways to finance services and pay physicians.

Private Health Insurance

There are three main categories of private health insurance: commercial insurance companies, Blue Cross and Blue Shield (the "Blues"), and employers who provide their own health insurance.

The "Blues" were originally developed as a not-for-profit insurance mechanism during the Great Depression of the 1930s to help hospitals stay financially viable by assisting unionized workers to purchase health insurance for their hospitalizations. In 1986, Congress had eliminated the nonprofit tax exemption, in response to large increases in premiums and retreats by the Blues from such cardinal principles as community rating.

The commercial for-profit insurance industry is made up of more than seven hundred companies, many of which are massive corporations with great political and economic influence by virtue of their interlocking directorates with major banks and corporations.

Many employers are developing their own insurance plans instead of purchasing those marketed by insurance companies. In 1985, 42 percent of all employees were covered by these self-insurance plans, compared with 5 percent in 1975. Self-insurance is attractive to employers because they can avoid the administrative costs of health insurance companies and maintain more direct control over the disposition of insurance payments. By 1985, self-insured employers had a greater share of the health insurance market than commercial insurers or the Blues. Another characteristic of self-insured plans is their association with "managed care"; by 1988, over 70 percent of individuals covered by employer-sponsored health insurance were enrolled in managed care plans, 18 percent in HMOs (Bodenheimer, 1990).

Tremendous diversity exists in premium structures and benefit packages of health insurance plans, which greatly increases administrative costs and partly explains the substantially higher costs of the U.S. health system compared with health systems in other countries. Table 8.5 provides an example of the diversity in types and extent of benefits in some of the health insurance plans in one typical state. It indicates that preventive care is not covered in most of the plans. In addition many of the plans contain provisions for a considerable deductible, which people must pay out of pocket for visits associated with illness and for diagnostic tests in the office. There are also other differences in coverage of prescription medication costs, general eye or dental care, surgeons' fees, and hospital care (Weiner and Frank, 1987).

Very few indemnity insurance plans cover all types of services. As a result, a large proportion of the population must pay out of pocket for important aspects of health care. While 91 percent of the population has some coverage for in-hospital care, only about 80 percent has at least some coverage for visits to physicians' offices, 88 percent for visits to hospital clinics, 80 percent for inpatient and 72 percent for outpatient mental health visits, 55 percent for skilled nursing facilities, 36 percent for home health care, 83 percent for maternity care, and 73 percent for prescription medicines. Only a minority of the population have even some coverage for dental care (25 percent), vision care (15 percent), routine preventive care (10 percent), and hearing problems (9 percent). Those whose coverage is through a government program (such as Medicare) are more likely to have coverage for the various types of services than those whose only coverage is through private insurance. (These data are from 1977, the last available.)

Even HMOs differ in the benefits they offer. For example, a 1988 survey indicated that 48 percent required copayments for primary care visits. The percentage of HMOs that covered preventive services, with or without a copayment, was as follows (Gold and Hodges, 1989):

Well-baby care	100%
Pap smears	100%
Diagnostic mammography	100%
Screening mammography	99%
Influenza shots	98%
Child immunization	99%
Adult immunization	97%
Routine physical	99%
Nutrition counseling	85%

| Health education classes | 76% |
| School/work physical | 36% |

Thus, private health insurance often fails to cover a substantial proportion of the costs of care, especially preventive care, so that patients have additional out-of-pocket expenses for services.

Government Financing and Provision of Services

Although the government has always assumed some responsibility for the medical care of individuals unable to pay for it, it was the Depression of the 1930s that resulted in a federal commitment on a large scale. Title V of the Social Security Act of 1935 and various amendments provided funds to states to support services for children with certain chronic conditions, maternal and child health clinics, family planning, regionalized prenatal care, and dental care. In 1965, the legislation was amended in a major way. For the first time, federal dollars were appropriated from general tax revenues to help states to pay for care for families with dependent children whose incomes were below a certain amount. By 1978 (the latest year for which information is known), public programs contributed 29 percent of the total expenditures for health care of children under age nineteen; two-thirds of the amount came from the federal government and one-third from states and localities. Just over half (55 percent) was direct reimbursement to physicians for services; the rest went to federally supported service programs (6 percent), the Department of Defense (14 percent), state and local hospitals (2 percent), and various other programs (3 percent) (Budetti et al., 1982). Four main types of programs are involved—Medicare, Medicaid, local health programs, and community health programs—as well as a variety of other smaller programs. A description of each follows.

MEDICARE

The almost universal coverage of the elderly for at least some health costs comes from the Medicare program, which resulted from congressional legislation in the mid-1960s. Medicare is a two-part program designed to help the elderly pay part of their medical bills. Part A is supported by a trust fund replenished each year by taxes on employee wages and employer payrolls. Individuals pay no premium for this governmental insurance other than the tax on their wages, but there are substantial deductibles as well as coinsurance after the first sixty days of hospitalization. Part B covers ambulatory care services and is financed by enrollee premiums (25 percent) and general tax revenues. It involves a monthly premium paid by enrollees and requires copayments. In addition there is an annual deductible sum. In mid-1988, the catastrophic care bill expanded Medicare to provide free hospital care after the annual deductible and limited out-of-pocket spending on part B. However, the legislation was subsequently repealed, in part because the elderly were unwilling to pay more taxes and additional premiums for the new benefits. Payment for preventive care is generally not covered by Medicare, but recent legislation (Omnibus Reconciliation Acts of 1990 and 1991) provided coverage for breast cancer screening and Pap smear examinations.

Policy Type	Preventive (Nonillness) Office Visit	Illness-Related Office Visit	Office Diagnostic	Prescription Medication	General Eye/Dental	Surgical Professional Fee	Hospital Facility Fee[b]
Blue Cross/Shield group (high option)	None	$100 Deductible, then 100% UCR	100% UCR up to $200, then 80% of UCR	$5.00 copay	Some	100% of UCR	In full
Blue Cross/Shield individual plan	None	$500 Deductible, then 80% of fee schedule	$200 Fee schedule coverage, then $500 deductible (then 80% schedule)	$5.00 Deductible, then 80% fee schedule	None	100% Fee schedule	In full (70 day max per episode)
Aetna Group (low option federal employee)	No adult, some well baby	$250 Deductible, then 75% of UCR	$250 Deductible, then 75% of UCR	$7.00 copay	Some	$250 Deductible, 75% of UCR	100% to $4,000. 75% thereafter. Stop loss of $6,000 patient payment
Traveler—individual ("Med-Pac")	None	Only hospital follow-up	None (except pre-admission)	None	None	100% of Fee schedule	Fee schedule (150 room & board. Up to $3,000 other charges)
Select care—PPO (individual)[c]	No adult, some well baby	$10 copay	Yes. Some copay	$200 Deductible, then 80% of costs	None	In full	$25/day copay for first 10 days. In full after that
HMO (care first) (state-group plan)	In full	In full	In full	$3.00 copay	Some	In full	In full

Source: Adapted from Weiner & Frank 1987

[a] UCR = usual, customary and reasonable. Deductibles are for each family member, individual and family "Stop Loss" limits may also apply. Maximum coverage varies.
[b] 365-day maximum unless noted otherwise.
[c] At PPO, use of "participating" providers assumed.

MEDICAID

The Medicaid program was initiated by federal legislation in 1965. Its intent was to assist states in paying for basic medical services for individuals who could not afford to pay either for care or for insurance against medical expenses. States determine the scope and design of their programs based on federal guidelines. The federal government matches state expenditures based on a formula depending on the state's wealth. States are permitted to exceed their matching requirement but must provide for a minimum of care, including inpatient and outpatient hospital care, physician services, laboratory and x-ray services, and nursing home care. States are permitted to establish their own income cutoffs for determining eligibility, except in the case of children and pregnant women. Medicaid is the only fee-for-service reimbursement mechanism that routinely pays for the preventive services that are important in primary care.

In the first decade of its existence, Medicaid eligibility and expenditures grew, but growth abated after 1976. Despite the economic recessions in the mid-1970s and early 1980s, with their high levels of unemployment and increasing poverty, the Medicaid population did not grow. States responded to budget pressure from the federal government by limiting income eligibility, reducing coverage of optional groups, and reducing the scope of covered services. Half of Medicaid recipients are children, although those under age twenty-one account for only 20 percent of Medicaid expenditures. From 1980 to 1985 Medicaid coverage of the poor and near-poor of all ages dropped from 53 to 46 percent of the eligible population. The 1985 average income eligibility standard for a family of four was $4,992, or 47 percent of the federal poverty level, with variability among states ranging from $1,776 per year or 21 percent of the federal poverty level in Alabama to $7,800 or 93 percent of the federal poverty level in California. The Omnibus Reconciliation Act of 1981 (OBRA) reduced federal funding and gave states increased flexibility to reduce eligibility standards; the nonelderly poor experienced the greatest loss of coverage, with the ratio of Medicaid enrollees to the poverty population declining from 52.1 to 45.5 percent. Of families then rendered ineligible for Medicaid, approximately one-half were left without any insurance coverage. A national survey of access to medical care showed that use of medical services declined markedly between 1982 and 1985, with needy children, especially those in poor health, suffering the greatest declines (Freeman et al., 1987).

In the early 1980s, other changes in the Medicaid program influenced its organizational format. Congress modified existing Medicaid procedures to permit prepayment for care of recipients without requesting special exemption from those provisions of the original Medicaid legislation that guaranteed freedom of choice of providers. By late 1985, there were a total of fifty-nine waivered programs in twenty-eight states. There are three main types of programs: mandatory choice, in which the recipient must choose a case manager from a list of specified providers; mandatory enrollment, in which the state contracts with an intermediary who is responsible for all covered services and may or may not allow the individual to choose a provider; and voluntary enrollment, in which the eligible individual may select a case manager or choose to remain in the fee-for-service system.

In these programs physicians can participate in different ways:

- They can act as the primary care provider, who assumes the responsibilities of a case manager and is paid a fee-for-service plus a small additional amount for management.
- The physician can contract with health insurance organization (HIO) in which primary care is arranged through an intermediary that negotiates payment rates. By law (1985) HIOs must maintain a non-Medicaid and Medicare enrollment of at least 25 percent, although some states have already received waivers to avoid this requirement.
- The physician can have prepayment contracts in which states contract with primary care organizations or physicians who become the case managers and are paid by capitation.
- Partially capitated services with physicians or organizations similar to the prepayment contracts, except that the capitation covers a more limited range of services and the case manager is not at risk for specialty services.
- Voluntary HMO enrollment in which the state maintains traditional Medicaid but encourages HMO development for Medicaid recipients.

HMOs serving Medicare and Medicaid patients have also been exempted from certain standards for operation of HMOs. These HMOs are called competitive medical plans (CMPs). The relaxation of standards for them is an effort by the federal government to increase the number of prepaid plans eligible to compete for publicly insured patients.

Although experimentation in the financing and delivery of Medicaid services continues to expand, there has been a rollback in the eligibility restrictions imposed in the early 1980s and, in the case of some subpopulations, certain expansions have occurred. The Deficit Reduction Act of 1984 (DEFRA) and the Consolidated Budget Reconciliation Act of 1985 (COBRA) required all states to provide Medicaid coverage to pregnant women and children up to age five in families with incomes below the AFDC payment level for a family of comparable size. This expansion of Medicaid eligibility required states to extend benefits to pregnant women and young children in two-parent families. Before this time, nearly half of the states excluded pregnant women and children regardless of their income if they did not meet certain family composition requirements.

The Omnibus Budget Reconciliation Act of 1986 (OBRA-86) permitted but did not mandate that states cover pregnant women, infants (initially and up to age four), and the elderly and disabled with incomes below the federal poverty level. By July 1987, at least half the states elected this option. The 1988 catastrophic care legislation mandated Medicaid coverage of pregnant women and infants in families with incomes under the federal poverty level, and this provision withstood the repeal of the general act. States now have the option to cover pregnant women and infants with incomes up to 185 percent of the poverty level and are required to cover children younger than six with incomes up to 133 percent of the poverty level. Effective July 1, 1991, coverage of children age 9 to 19 with incomes below poverty is required and will be phased in by ages over the next ten years.

LOCAL HEALTH DEPARTMENT PROGRAMS

The third type of public assistance program involves local health departments. Their role has waxed and waned. In the early decades of the century, public health efforts to provide both environmental and direct maternal and child health services had a major impact in reducing infant and early childhood mortality. In the 1950s, population migration, particularly from southern rural areas to large municipal areas, found local health departments unprepared; as a result, the earlier declines in rates of infant mortality ceased. During the 1960s, the domestic War on Poverty rekindled commitment to public health in the form of community health clinics, supported by direct federal grants as well as by state and local dollars. The wave of privatization and marketplace competition in the early 1980s resulted in declining roles for local health departments in providing of direct health services. Today, efforts are concentrated largely on environmental health matters and communicable disease control. All states now require children to have basic immunizations as a condition of entry into school and organized day care programs. Some also require a physical examination for entry into school and periodically thereafter, and some jurisdictions provide school health services where children are not otherwise able to obtain them.

Almost all (forty-six) states have local health departments. Most report that they provide immunizations, surveillance of reportable diseases, and child health services. Half to three-quarters are involved in health education, prenatal care, AIDS testing and counseling, chronic diseases, home health care, and services for handicapped children. Fewer (35 to 49 percent) provide some laboratory or dental services. Less than one-quarter have any program in occupational safety and health, primary care, obstetrical care, drug and alcohol abuse, mental health, emergency medical services, long-term facilities, or hospitals (MMWR, 1990). Consequently, local health departments do not play a major role in primary care, although they may assume responsibility for some components of primary care if these are not being adequately provided elsewhere.

COMMUNITY HEALTH CENTERS

Initiated in the mid-1960s as an effort to provide comprehensive health services for underserved populations, these centers have grown into a network of 540 facilities in 2,000 locations serving nearly 6 million poor and underserved individuals in 50 states, Puerto Rico, the Virgin Islands, Guam, and the District of Columbia. The centers are funded primarily through grants from the federal government through the Community Health Centers, Migrant Health Centers, and Homeless Health Care programs. About half of all health center physicians are loan and scholarship recipients fulfilling their obligations under the National Health Services Corps Act. Health centers by law must serve all Medicaid and Medicare patients as well as those without health insurance. However, only one-fifth of the country's 30 million medically underserved are reached by existing facilities.

OTHER HEALTH-RELATED PROGRAMS

Although maternal and child health (MCH) services have been provided through federal grants for over five decades, the new federalism policies in the early 1980s resulted in the formation of block grants that consolidated funding to the states. The program contains grants for maternal and child health services, Supplemental Security Income for Disabled Children (for children with chronic health problems), lead-based paint poisoning prevention programs, sudden infant death programs, hemophilia treatment centers, and adolescent pregnancy and genetics services grants. In most southern, midwestern, and western states that have large networks of county and local health departments, services are often provided directly through government agencies in schools, local health clinics, well-child clinics, antenatal clinics, health screening, and immunization services. In other areas, particularly the northeast, the state MCH agency often contracts with facilities in the private sector to provide services.

In 1975, Congress passed the Education for the Handicapped Act (Public Law 94-142) to provide federal funding for state education departments to ensure free public education for handicapped children of age three to twenty-one. More recent legislation (Public Law 99-457) extended the program to children from birth to three years of age. Approximately 10 percent of children, including those with deficits such as hearing problems, are served across the country, but coordination between the educational services and health services for these children is inadequate.

The Women, Infants, and Children (WIC) program was initiated in 1972 to provide supplemental food and nutrition education to pregnant and postpartum women, nursing mothers, and children from birth to age five whose families incomes are below 185 percent of the federal poverty level. The program is administered by the Department of Agriculture through grants to state health departments. Only about half of all eligible women, infants, and children receive WIC services.

The Head Start program resulted from legislation in the mid-1960s designed to improve the developmental level of low-income preschool children. It is administered by the Federal Administration for Children, Youth and Families, with support given directly to local Head Start groups. Although primarily directed at enhancing the likelihood of success in school, health goals were articulated from the program's inception.

Support for family planning comes from a variety of public and private sources, including Title X of the Public Health Service Act, Title XX of the Social Security Act, Medicaid and community health centers, and the MCH block grant. Services are provided through specialized clinics, local health departments, community health centers, hospitals, and private physicians' offices.

Payment of Physicians

By 1986 almost half (48 percent) of all physicians participated in at least one prepayment plan and nearly one-fifth (18 percent) of office-based physicians worked with three or more such organizations. However, only a small percentage of these doctor's visits were prepaid, even for physicians affiliated with HMOs. Physicians with HMO affiliations typically received 10 percent of their practice earnings from prepaid plans, with the remainder coming from traditional fee-for-service. Pediatricians, obstetricians/gynecologists, and family physicians had the highest proportion (12 to 13 percent), while surgeons (5 to 8 percent, depending on type of surgeon), internists, and psychiatrists had the lowest proportion (9 percent) of earnings from prepayment (Owens, 1987). Therefore, despite the dramatic increase in HMOs and prepayment plans, fee-for-service continues to be the primary form of payment of physicians in the United States.

Arrangements for payments to physicians through more than one mechanism are becoming the norm in the United States. In 1987, about 40 percent of all physicians engaged in patient care but not employed by the national government had at least one contract with a preferred provider organization (PPO). This proportion was much greater than just four years earlier when it was about 11 percent (Emmons, 1988). Twenty-five percent of physicians were members of an independent practice association, and 35 percent had at least one contract with an HMO. About 6 percent of physicians received at least 90 percent of their income from an IPA or HMO; 12 percent received none of their income from them. The remainder of physicians received varying proportions of their income from either IPAs or HMOs, the plurality (31 percent) receiving between 1 percent and 5 percent of their income from them. Physicians in the western region of the country and those just beginning in practice (under thirty-five years of age), as well as pediatricians, were more likely to receive larger proportions of their income from IPAs or HMOs. Fewer than 1 percent of physicians received most of their income from PPOs; about 12 percent received none of their income from a PPO and about 45 percent between 1 and 5 percent, with little difference by geographic area, age, setting, or specialty (Emmons, 1988). The most common combinations of sources of income, at least among pediatricians, were noncontract fee-for-service and contract fee-for-service (11 percent) and noncontract fee-for-service and salary (8 percent). None of the other combinations such as contract fee-for-service, noncontract fee-for-service, salary, or some other source, characterized the income sources of more than 4 percent of pediatricians (LeBailly, 1985, pp. 16–19).

In the United States the income of specialists is approximately double that of primary care physicians (Gonzalez and Emmons, 1989), and specialists are paid more than generalists even for the same procedure (Weiner and Frank, 1987). This results from a system of payment that rewards the performance of tests and procedures. The historical approach to payment is based on the "usual and customary fees" charged by individual physicians in particular areas of the country. Newly available technology is generally priced at relatively high levels. Relative value scales, which were intended to reflect the prices of one procedure compared with others, formalized the disparities between reimbursements provided for technology and interpersonal interventions. Since specialists are more likely than primary care physicians to employ these technologies, their income from reimbursements rapidly escalated compared with those of primary care physicians, who generally perform less costly procedure and rely more heavily on interpersonal interventions such as medical counselling. The wide disparity between the incomes of primary care physicians and those of specialists is likely to be reduced in the foreseeable future as a result of physician payment reforms mandated by the U.S. Congress. Although these reforms initially will apply only to the Medicare program, they may be adopted by other third-party payers or by managed-care organizations as well. Two approaches are under consideration. In the first, reimbursement for procedures which are overpriced (based on their actual costs compared with existing prices) will be reduced. In the second approach, increases from year to year in the Medicare economic index will be greater for primary care than for specialties, thus reducing existing disparities in payment between primary care and specialty services.

TABLE 8.6.   Estimated Percentage of Population Covered by Health Insurance
by Type of Health Plan

| | | Type of Health Plan | | | |
| | Percent of | Traditional | Managed | | |
Type of Insurance	Population	Indemnity	Indemnity	PPO	HMO
Private					
1. Employment-Linked					
Self-insured	34	10	50	20	20
Not self-insured	27	30	35	15	20
2. Individual	8	70	15	10	5
All private	69	25	35	20	20
Government					
1. Medicare	10	96	0	0	4
2. Medicaid	8	80	10	5	5
All government	18	85	8	2	5
Uninsured	13	NA	NA	NA	NA
Total	100				

Source: Adapted from Weiner and deLissovoy, 1991 (submitted for publication).

Access to Health Services

Except among the elderly, basic health insurance is inequitably distributed in the population. A substantial minority of the population of the United States has no assured financial access to services. In 1989, 16 percent of the population had no external source (third-party coverage) of funding. The proportion without third-party coverage was greatest for children under age five, 17 percent of whom had no source of funding other than out-of-pocket coverage. By ages forty-five to sixty-four, about one in ten adults (10.6 percent) had no third-party coverage. For those over age seventy-five, however, more than 98 percent of individuals had some source of help with payment of medical charges (Health US, 1990, pp. 208–9). Males are slightly less likely to have no coverage than females, and there are marked differences according to family income: 37 percent of those in the lowest income group and 21 percent in the second lowest income group have no coverage at all, compared with 3.2 percent in the highest income group and 5.6 percent in the next highest group. The source of coverage also varies by family income; 96 and 92 percent of the top two highest income groups but only 35 and 71 percent of the two lowest income groups have some private (nongovernmental) insurance (Health US, 1990, p. 208). Table 8.6 shows the wide diversity in source and type of health insurance in the population.

In 1989, the average number of contacts with physicians per person in the United States was 5.3, over half (60 percent) took place in a doctor's office, 13 percent in a hospital outpatient department, and 12 percent by telephone, with the remainder in out-of-hospital clinics, laboratories, or the home. On average females had more contacts than males: 5.6 versus 4.4. The number of contacts per person was greatest in families with low income (5.8), but those in the second lowest income group, who are least likely to have health insurance or governmental assistance with health care costs, averaged only 4.8 contacts per year compared with 5.5 in the highest group and 5.2 in the next highest group. The type of care also differed by family income. For those in the lowest income group, 49 percent were in a doctor's office, 20 percent in a hospital clinic, and 11 percent by phone, compared with 63 percent, 11 percent, and 14 percent, respectively, for those in the highest income group (NCHS, 1991).

Although the overall mean number of physician visits per person decreased through the 1980s, visits by poor or near-poor individuals declined even more than the average. The disadvantage was even greater for those who were in fair or poor health; the gap in mean number of visits between the poor or near-poor who were ill and the nonpoor ill widened from 12 percent more in 1982 to 27 percent less in 1986. Another trend in medical care is the decreasing proportion of the population who have a regular source of care; in 1986, 18 percent of the population, compared with 11 percent in 1982, reported no single usual source of care (Freeman et al., 1987).

The importance of unimpeded access to health services cannot be overemphasized. Health improves when barriers to access are reduced (Starfield, 1985; Roemer, 1984). Conversely, health status worsens when families are required to pay a coinsurance fee for their medical services. When the state of California imposed a copayment by requiring $1 per visit from certain Medicaid beneficiaries, the frequency of doctor visits fell compared with prior levels and with beneficiaries not subjected to the requirement. After a lag period, hospitalization rates in the copay group rose, both absolutely and compared to the group requiring no payment. The increase in hospitalization rates, a likely result of neglect of primary medical care due to the inhibiting effect of the copayments, resulted in costs which more than offset the savings to the program generated by the copayments (Roemer et al., 1975).

In a large-scale insurance experiment, families were randomly assigned to one of a variety of plans differing in the rate of coinsurance for medical visits, one plan requiring no fee at all. Both health status and use of services were monitored for several years. As compared with the free plan, both adults and children in the copayment groups were less likely to receive care for conditions for which medical care is effective. The adverse effect of coinsurance was even more striking for the financially needy than it was for others (Lohr et al., 1986, p. S36). Needy individuals in the coinsurance groups also were in poorer health at the end of the experiment than comparable individuals in the group receiving free care. In all groups, requirements for copayment reduce equally the use of needed and discretionary or inappropriate services (Siu et al., 1986).

Copayments also reduce use of services in health maintenance organizations; reductions in use are more marked for primary care than they are for specialty services (Cherkin et al., 1989).

The Future of Medical Practice Organization

Medical practice in the United States is likely to undergo radical change in the near future. Recent directions suggest an increasing corporatization of medicine, with the ownership of facilities and control of medical groups being assumed by large for-profit companies that contract with physicians and physician groups to provide care. Concern about expenditures for care are resulting in increasing scrutiny of medical practice and the imposition of various types of control over it. Both community-based and hospital-based physicians will be affected; assumption by corporations of ownership of hospitals, including teaching hospitals, is occurring at an increasing rate (Whiteis and Salmon, 1987).

The trend toward managed care has several implications for the practice of medicine. As employers, especially large corporations, and insurance companies increasingly assume responsibility for managing care, the practice of medicine will increasingly become a matter of renegotiated contracts. Managers, whose emphasis is likely to be on cost control, will seek new sources of care that are less costly, and individuals insured under these "managed care plans" may be subjected to periodic changes of physicians. These managed systems of care also often substitute other types of personnel for physicians for certain types of services. Although the emphasis on cost containment is likely to reduce access to health services and possibly decrease the scope of services covered by insurance, managed health systems could contribute to studies of the quality of care. To maintain control over services, they must maintain information systems on enrolled populations. These information systems are potentially useful in research. For example, little is known about the extent to which commonly employed procedures actually achieve the purposes for which they are intended. The information systems maintained in managed care system could be used for medical effectiveness studies, much as the Medicare claims files have been used in studies of the effectiveness of health services for the elderly.

This general overview indicates the wide variety of arrangements for providing care and the diversity in sources of payment and coverage for various types of services and their impact on access to care. This extraordinary organizational and financial disarray of a health services system is virtually unique in the industrialized world. As a result of the disarray, those who need care often lack the necessary information to decide where to obtain it and how to pay for it. The fragmentation of organizational and financial arrangements leads to costs that are much higher than those in better organized systems of care, even those that are still primarily in the private sector, such as in Canada (Woolhandler and Himmelstein, 1991). Will more rationality be brought into the U.S. system of health care to improve access to and organization and financing of primary care? The answer depends on whether or not the prevailing emphasis on marketplace competition can be tempered by concerted efforts to reduce the disadvantages of the current system and to maximize the potential of the new forms of organization for increasing rather than decreasing access to services of high quality.

References

Bodenheimer T. Should we abolish the private health insurance industry? Intl J Health Services 1990; 20:199–220.

Budetti P, Butler J, and McManus P. Federal health program reforms: Implications for child health care. Milbank Memorial Fund Quarterly/Health & Society 1982; 60:155–81.

Cherkin D, Grothaus L, and Wagner E. The effect of office visit copayments on utilization in a health maintenance organization. Medical Care 1989; 27:1036–45.

Clancy C and Hillner B. Physicians as gatekeepers: the impact of financial incentives. Arch Int Med 1989; 149:917–20.

Dalton JJ. HMOs and PPOs: similarities and differences. Topics Health Care Finance 1987; 13:8–18.

deLissovoy G, Rice T, Gabel J, and Gelzer H. Preferred provider organizations one year later. Inquiry 1987; 24:127–35.

Eisenberg JM and Kabcenell A. Organized practice and the quality of medical care. Inquiry 1988; 25:78–89.

Emmons DE. Changing dimensions of medical practice arrangements. Medical Care Rev 1988; 45:101–28.

Freeman H, Blendon R, Aiken L, Sudman S, Mullinex C, and Corey C. Americans report on their access to care. Health Affairs 1987; 6:6–18.

Gold M and Hodges D. Health maintenance organizations in 1988. Health Affairs 1989; Winter:125–38.

Gonzalez M and Emmons D. Socioeconomic Characteristics of Medical Practice 1989. AMA Center for Health Policy Research, American Medical Association, 1989.

Gruber L, Shadle M, and Pion K. The InterStudy Edge 1989; 4:1.

Havlicek P. Medical Groups in the U.S. A Survey of Practice Characteristics. Chicago, American Medical Association, Division of Survey and Data Resources, 1990.

Hillman AL, Pauly MV, and Kerstein JJ. How do financial incentives affect physicians' clinical decisions and the financial performance of health maintenance organizations? New Engl J Med 1989; 321:86–92.

Kasper JD. The importance of type of usual source of care for children's physician access and expenditures. Medical Care 1987; 25:386–98.

LeBailly SA. A Profile of the American Academy of Pediatrics' Fellows: Descriptive Data from the AAP Membership Census. Statistical Note # 3. Division of Health Services Research and Information. American Academy of Pediatrics, Nov 1985.

Lohr KN, Brook RH, Kamberg CJ, Goldberg GA, Leibowitz A, Keesey J, Reboussin D, and Newhouse J. Use of medical care in the Rand Health Insurance Experiment. Diagnosis- and service-specific analyses in a randomized controlled trial. Medical Care 1986; 24 (Suppl) (9):S1–S86.

Luft H. Health Maintenance Organizations: Dimensions of Performance, New York, John Wiley & Sons, 1981.

Mortality and Morbidity Weekly (MMWR). Selected characteristics of local health departments, United States, 1989. 1990; 39:607–10.

National Center for Health Statistics (NCHS), AJ Moss and VL Parsons. Current Estimates from the National Health Interview Survey, United States, 1985. Vital and Health Statistics. Series 10 No. 160. DHHS Pub No (86-1588). Public Health Service. Washington, D.C., U.S. Government Printing Office, Sept. 1986.

National Center for Health Statistics (NCHS) Health United States 1990. Hyattsville, Md., Public Health Service, 1991.

Owens, A. What's prepaid care worth to doctors? Medical Economics 1987; 64:202–19.

Retchin S and Brown B. The quality of ambulatory care in Medicare health maintenance organizations. Am J Public Health 1990; 80:411–15.

Roemer M, Hopkins C, Carr L, and Gartside F. Copayments for ambulatory care: penny-wise and pound-foolish. Medical Care 1975; 13:457–66.

Roemer M. The value of medical care for health promotion. Amer J Public Health 1984; 74:243–48.

Siu AL, Sonnenberg FA, Manning WG, Goldberg GA, Bloomfield ES, Newhouse JP, and Brook RH. Inappropriate use of hospitals in a randomized trial of health insurance plans. N Engl J Med 1986; 315:1259–66.

Starfield B. Effectiveness of Medical Care: Validating Clinical Wisdom. Baltimore, The Johns Hopkins University Press, 1985.

Starr P. The Social Transformation of American Medicine. New York, Basic Books Inc., 1982.

Stoline A and Weiner J. The New Medical Marketplace: A Physician's Guide to the Health Care Revolution. Baltimore, The Johns Hopkins University Press, 1988.

Weiner J and de Lissovoy G. Razing a tower of Babel: A taxonomy for managed care and health insurance plans. J Health Polit Policy Law 1993; 18(1):75–103.

Weiner J and Frank R. An analysis of the impact of private insurance on the delivery of primary care in Maryland. Report prepared for the Office of Policy Analysis and Program Evaluation. Maryland Department of Health and Mental Hygiene, April 1987.

Welch W, Hillman A, and Pauly M. Toward new typologies for HMOs. Milbank Mem Fund Q 1990; 68:221–43.

Whiteis D and Salmon J. The proprietarization of health care and the underdevelopment of the public sector. Internl J Health Serv 1987; 17:47–64.

Woolhandler S and Himmelstein D. The deteriorating administrative efficiency of the U.S. health care system. N Engl J Med 1991; 324:1253–58.

B

Issues in Primary Care

9

What Types of Physicians Should Provide Primary Care?

Primary care in the United States differs from that in most other countries in that it is provided not only by family physicians but also by general internists (for care of adults) and by pediatricians (for care of infants and children). There is also evidence that some specialists provide primary care, at least part of the time (Aiken et al., 1979). Two questions arise: Is the practice of primary care equally effective, regardless of the type of physician? Or does theoretical or practical justification exist for limiting the practice of primary care to certain types of physicians?

The Theoretical Basis for the Primary Care Physician

Dr S is a well-known physician and biomedical researcher who began to lose his vision in his early sixties. Given his background as a physician and distinguished scientist, he had no problem in searching out the best ophthalmologists in the field. Nor did he have trouble communicating with them. He spoke their language and shared their knowledge. Dr. S was diagnosed as having macular degeneration, an irreversible and progressive cause of blindness, and was told there was nothing that could be done for him. Only much later, and through discussions with nonprofessionals, Dr. S learned that much could be done for him. These are his words:

"I think it is the rare physician who has the time and energy and insight to devote profound attention to what is happening to the rest of his patient. It may be that my physician can do nothing for my eyeballs but behind those eyeballs is an anxious, worried man who demands inputs but is now deprived of his accustomed inputs, and there is lots that can be done for him. The complications of blindness are not in the eye but elsewhere. One of them is a feeling of soreness on the anterior surface of both tibias about 12 inches from the floor, which is about the height of the conventional coffee table. It is a problem of a patient who is blind, but it isn't mentioned in any of the ophthalmological textbooks. I believe that blindness to most ophthalmologists represents failure, just as death represents failure to the internist. The internist doesn't have to deal with the patient who has died, but the patient who is blind lives on, and the inability of the ophthalmologist to cope with the blindness has a major impact on the patient. Ophthalmologists deny the patient any other help because they feel defeated. Ophthalmologists are unaware of dozens of aids that make life more tolerable for the blind patient.

"Had my doctor told me about talking wristwatches and talking books, my world would have been transformed from a living hell to a roaring inferno, and sometimes to a heaven."

Dr. S went to the "best" ophthalmologist available by all professional criteria. He might have been better off being seen by a primary care physician, with consultation from the ophthalmologist.

From a theoretical viewpoint, the interests and skills involved in providing primary care should be different from those involved in specialist care. Physicians who practice primary care must tolerate ambiguity since many problems never reach the stage of a diagnosis that can be coded with standard diagnostic nomenclature. They must be comfortable in establishing and maintaining relationships with patients and in dealing with problems for which there is no demonstrable biological aberration. They also must be able to manage several problems at once, even though the problems may be unrelated in etiology or pathogenesis. Furthermore, over long periods of time, the patients' problems change, and specialists who were appropriate for the original problem may be challenged beyond their skills and interests by other types of problems that arise.

Medical progress and new technologies provide the impetus for an increasingly specialist orientation and also demand increasingly well-honed generalist skills. Better strategies for management will improve life expectancy, increasing the complexity of illness management over time; increased survival will also result in the accumulation of different types of illnesses within individual patients. Health problems are becoming more complex, with more syndromes, handicaps, impairments, and disabilities than those to which physicians are accustomed. At one end of the life cycle will be an aging population and at the opposite end a greater proportion of handicapped survivors of neonatal intensive care. Both developments are likely to heighten the need for home-based care and for knowledge about the existence of community resources.

These considerations suggest that primary care should be provided by individuals trained for primary care in primary care settings rather than by those trained in tertiary medical centers.

The theoretical argument for primary care physicians assuming responsibility for primary care is buttressed by experience in other professions, in the health systems of other industrialized nations, and in organized health care systems.

As Moore (1990) noted, most other professions and industries employ specialization to deal with increasing complexity. However, when size and complexity reach a certain level, organizations employ general managers who plan, allocate resources, supervise and coordinate the work of specialists, and monitor the results. Few industrialized nations rely on specialists to provide primary care; most have systems that are based on the generalist physician, with backup from specialists (Kaprio, 1979). In the United States, all organized health services systems, such as group practice forms of health maintenance organizations, employ primary care physicians; Moore (1990) estimated that in the late 1990s virtually all the graduates of primary care programs would be needed to staff the HMOs if their programs grow, as currently predicted, at a rate of 10 percent per year.

Empirical evidence of the benefits of a primary care physician supports the theoretical notions. The higher the proportion of general physicians in a community, the lower its frequency of hospitalization. Experiences in Sweden indicate that primary care both reduces the flow of patients into specialized secondary care services such as consultants and emergency rooms and lowers age-adjusted total health care costs (Moore, 1990). Data from an international collaborative study of medical care utilization showed that areas with higher specialist-to-generalist ratios have higher rates of physician visits, which cannot be accounted for by greater health needs (Kohn and White, 1976). Specialists may overtreat patients; for example, allergists treating children and adults with asthma use more corticosteroid than do the family physicians and pediatricians with whom they have been compared (Engel et al., 1989).

If training specifically in primary care is important for its practice, what evidence is there that certain types of such training are superior to others? In particular, is the training of family physicians, general internists, and pediatricians equally effective? The research literature provides a wealth of data on this issue. Almost all studies considered here were conducted in the United States, the only Western industrialized country in which general internists and pediatricians have equal status as primary care physicians with family physicians or "generalists."

The literature contains several approaches to comparing the practice of family physicians and general internists in the care of adults with that of family physicians and pediatricians in the care of children. (A few studies have also included specialists of different types in their comparisons.) Diagnostic and treatment methods, referral practices, and the use of resources have been studied by a variety of methods including medical record reviews, logs kept by physicians for several days, questionnaires sent to physicians, and programmed (simulated) patients. Some studies have been national in scope whereas others were limited to individual clinical facilities. Some just describe differences in practice characteristics, whereas others have assessed the technical quality of care, satisfaction with care on the part of patients, reductions in the utilization of services, or some aspects of the outcome or costs of care. A few studies have addressed the attainment of the unique feature of care; these studies are summarized in Chapters 3 through 6.

Profiles of Care: Descriptions of Physicians' Practices

The National Ambulatory Medical Care Survey, a periodic survey in which a random sample of physicians keep a log of patients for several days, provides descriptive information on a variety of characteristics of patients, their problems, and practice characteristics. This survey can be used to compare the practices of various types of physicians. For example, internists spend more time with patients (18.4 minutes) than family–general practitioners (13.0 minutes). Internists order more laboratory tests (73 percent of visits) and x-ray tests (53 percent of visits) than generalists (34 and 19 percent, respectively). Internists are also more likely to provide instructions regarding health problems (17.8 percent of visits) than generalists (12.4 percent), but there are no apparent differences in frequency of therapy for emotional problems (3 percent; Noren et al., 1980).

The same survey has been used to compare characteristics of the practices of family physicians and general practitioners with those of pediatricians. Pediatricians order more laboratory tests but prescribe fewer drugs for the major categories of illnesses (fever, sore throat, abdominal pain, diarrhea, and earache) (Fishbane and Starfield, 1981).

Logs kept by physicians were the source of information in another national survey of face-to-face encounters of family physicians and pediatricians. One study from this survey examined the care provided to children in five types of visits (well-patient, skin, ear, mouth or throat conditions, upper respiratory conditions) and with five major diagnoses (medical examination, upper respiratory infections, pneumonia, pharyngitis, otitis media) in the ambulatory setting. Pediatricians performed more diagnostic tests for all diagnoses, gave more immunizations, and offered more counseling about growth and development, but they provided less specific therapy than family physicians. However, family physicians did more counseling about family and sex matters than pediatricians (Table 9.1). Family physicians were also more likely to provide a broader range of services, especially minor surgery, for every age group in childhood (Starfield et al., 1985).

The same national survey was used to study a sample of 132 family physicians and 102 general internists who completed questionnaires and kept log diaries on 3,737 and 2,250 adult office visits, respectively. Analysis revealed that general internists were twice as likely as family physicians to order blood tests, blood counts, chest x-rays, and electrocardiograms. They also spent more time with patients, and referred and hospitalized them at a greater rate. The average charge per visit for patients of internists was about twice that for patients of family physicians, even after controlling for a variety of patient, practice, and physician characteristics (Cherkin et al., 1987).

Diagnostic methods of internists and family physicians have been compared by employing research assistants trained to present problems as if they were real patients. For example, the diagnostic methods of nine family physicians were compared with those of nine internists, using these "programmed" patients to present three clinical problems. Family physicians asked fewer history questions, requested fewer items of data about physical examination, and ordered fewer laboratory investigations. In two of the three problems, the study revealed that family physicians asked relatively more questions about mental status and life situation. There were no significant differences in the diagnoses reached (Smith and McWhinney, 1975).

In another study using simulations, the diagnostic methods of third-year residents in internal medicine ($n = 31$) were compared with those in family practice ($n = 22$) using written descriptions of five patients, each with a different problem. The family practice residents targeted far fewer physical examination items. Laboratory charges were greater for the internists, but only for two of the simulations. However, the number of diagnostic hypotheses presented by the two groups of physicians did not differ (Scherger et al., 1980).

Medical records were the source of information in other comparisons of the care provided by different types of primary care physicians. In one such study, 520 patients were randomized to an internal medicine or a family practice clinic. After

TABLE 9.1. Percentage of Visits of Different Type by Type of Physician and Age of Child. Outpatient Child Visits, United States, Mid-1970s

Age of Child and Type of Visit	Type of Physician					
	Family Pediatrician n = 21,926	General Physician n = 5,772	Practitioner n = 3,698	Other Specialist n = 11,270		
Less than 1 year						
Preventive	51.5	44.0	58.4	18.4		
Medical	46.9	53.1	40.4	50.7		
Minor surgery	0.9	1.4	0.3	11.9		
Major surgery/ Med-surgical	0.6	1.1	0.4	18.5		
1 to 4 years						
Preventive	25.1	17.9	21.6	11.6		
Medical	71.5	74.0	67.8	44.6		
Minor surgery	2.6	6.1	5.4	18.7		
Major surgery/ Med-surgical	0.6	1.2	0.9	23.7		
5 to 9 years						
Preventive	16.4	10.9	22.8	11.9		
Medical	77.5	78.0	62.2	44.4		
Minor surgery	4.1	8.6	12.5	19.2		
Major surgery/ Med-surgical	0.5	1.0	1.5	23.6		
10 to 14 years						
Preventive	21.1	11.5	30.1	13.5		
Medical	70.3	64.7	55.9	48.4		
Minor surgery	5.3	15.3	11.1	14.8		
Major surgery/ Med-surgical	1.1	6.8	2.2	20.1		
				Internist n = 622	OB-gyn n = 1,895	Other n = 10,398
15 to 19 years						
Preventive	21.8	11.5	23.4	13.8	16.7	11.6
Medical	67.6	63.1	60.6	74.1	21.8	45.7
Minor surgery	6.2	12.7	10.0	7.1	7.1	12.7
Major surgery/ Med-surgical	0.7	3.1	1.1	2.7	3.2	13.4

about two years, charts of the patients were evaluated for frequency of visits, laboratory studies ordered, number of referrals, acute care clinic visits, emergency room visits, and frequency of broken appointments. Associated costs were also ascertained. Patients seen by the internists, on average, made more frequent visits to the primary care clinic, the emergency room, and the acute care clinic. They were less likely to have kept their appointments with the primary care clinic. The median total annual cost of laboratory tests for patients in the internal medicine clinic was significantly higher because of higher referral rates to specialists and higher laboratory test charges generated by specialists (Bertakis and Robbins, 1987).

Medical records were also used to study about 2,000 inpatients of family physicians and internists regarding their length of stay, charges generated, charges generated per day, disposition, number and type of diagnoses, and number of procedures. The only differences were in the number of diagnoses; family physicians assigned fewer. Review of a random sample of fifty charts of family physicians and a matched sample from internists revealed no differences in severity of illness, and multivariate adjustment for differences in case mix did not change the findings (Franks and Dickinson, 1986).

Surveys of physicians can also provide information on comparative practice patterns. In one such survey, a representative sample of physicians in Maryland were queried about their attitudes toward patients with hypertension. Family physicians were slightly less likely to indicate that they would order certain diagnostic tests for patients. They were also more cautious in recommending antihypertensive medication with mildly increased blood pressure but were more likely to favor some nonpharmacologic regimens (reducing weight and smoking, exercising, reducing alcohol intake, and adopting a low-cholesterol diet) than either internists or other specialists. All types of physicians were equally likely to support the use of diuretics for initial therapy of patients with mild hypertension (Cloher and Whelton, 1986).

In another survey, a questionnaire asked family physicians, internists, and obstetrician/gynecologists in Maryland to report the percentage of their patients who had been referred and the percentage whom they referred. Family physicians received fewer referrals than did the other two types of physicians, but referred the same percentage (10–11 percent) of patients as did generalists (Sobal et al., 1988).

Patients have also been queried concerning their physician's practice patterns. In one such study, college freshmen were asked about the extent of health counseling they had received from their primary health care providers. Patients of internists had received more counseling about smoking and alcohol use than patients of pediatricians or family physicians, more counseling about drug abuse and heart disease prevention than patients of pediatricians, and more counseling about weight control and nutrition than patients of family physicians. Pediatricians gave more counseling on weight control than family physicians but did not exceed internists in any area mentioned here (Joffe et al., 1988).

These studies of profiles of care indicate that pediatricians and internists do more extensive diagnostic workups than family physicians. Is there any evidence to suggest that this is associated with less satisfaction, poorer diagnosis, management, or outcomes of care? The following categories of studies address these questions.

Patients' Satisfaction with Care

Satisfaction of patients with their care has been studied by follow-up telephone interviews as well as mailed questionnaires. In one study using both medical records and telephone interviews, satisfaction was greater among family practice patients than with a matched group of patients seen by internists or pediatricians (Farrell et al., 1982).

Questionnaires sent to patients provided information from a national study in which patients were randomly sampled from logs kept by family physicians and internists. These patients were asked how satisfied they were with the medical care they were receiving. Two hundred thirteen adult patients of 124 family physicians and 218 patients of 98 general internists responded. Patients of both types of physicians reported similar levels of satisfaction on all four dimensions measured (access, humaneness, quality, and general satisfaction) even after controlling for a variety of patient, practice, and physician characteristics (Cherkin et al., 1988).

Technical Quality of Care

Physicians' recognition of patients' problems and their adherence to professionally defined standards of care for diagnosis and management in primary care have been studied by a variety of approaches including medical record review, physician interviews, and auditing of claims forms.

Audits of medical records were the basis of a study that found family physicians recorded fewer of the health supervision criteria than pediatricians but did as well in recording items related to disease management (Thompson and Osborne, 1976).

When quality of care was judged by examining medical records to ascertain diagnosis and management for specific conditions, specialists performed better in their own area of specialized training than did family or general practitioners or specialists performing outside their specialty areas (Payne et al., 1984). This is consistent with the known relationships between volume of care and quality of care on inpatient units in hospitals; the higher the volume of care for a problem by a particular surgeon or by a hospital in general, the better the quality of care. However, the *overall* quality of care by a given practitioner or particular practice cannot be judged by assessing the care for any particular diagnosis, especially when the selected diagnosis reflects problems often seen in specialty care rather than those more characteristic of primary care.

In another study, a screening test was given to 1,452 patients attending two primary care clinics in a large comprehensive health care facility to identify existing psychosocial problems. Family physicians were less likely than internists to recognize existing mental health problems, but the research design could not determine if the differences were due to other differences in practice style in the two clinics or to the difference in type of physician (Kessler et al., 1985).

In another study, both medical records and interviews of primary care physicians were used to study the attainment of a minimum set of standards (developed by physician consensus) for managing patients with four indicator conditions: care of normal infants, care of pregnant women, care of adult onset diabetes, and care of patients with congestive heart failure. A random sample of all physicians in one county were asked to participate; the sample included 34 family physicians, 11 internists, 8 pediatricians, and 8 obstetricians with 523 infants. It involved 363 pregnant women, 244 diabetic patients, and 128 patients with congestive heart failure who visited the practices during the time of the study. Data were collected from medical records and interviews with physicians. Management scores were better for pediatricians and obstetricians for two conditions—care of normal infant and pregnant women, respectively—than for family physicians. For the other two conditions—adult onset diabetes and congestive heart failure—no differences existed between the management scores of family physicians and those of internists.

Neither patient nor disease characteristics explained the differences that were found (Hulka et al., 1976).

A total of 190 physicians responded to questionnaires sent to 200 randomly chosen pediatricians and 300 family physicians in Georgia inquiring about their treatment of enuresis, their prescribing habits, and their knowledge of the side effects and toxic effects of tricyclic antidepressants. Nearly half of both family physicians and pediatricians reported using tricyclic antidepressants to manage enuresis. Only one-third of prescribers of these drugs were aware of their side effects or toxicity. Among physicians who did *not* use the drug, a greater proportion of pediatricians were aware of side effects and toxicity than nonusing family physicians. Pediatricians were more likely to treat patients for a shorter period of time, to limit prescription refills, and to promote the availability of ipecac syrup in the home for use in an accidental ingestion (Rauber and Maroncelli, 1984).

In an evaluation of a state peer review system, audited claims forms revealed that general practitioners had a smaller proportion of injections denied by the peer review organization than internists but more than pediatricians. However, general practitioners tended, on average, to give more injections per visit and had higher absolute numbers of injections denied per visit than either internists or pediatricians had. On average, general practitioners tended to give more injections per visit than other types of physicians and had slightly more injections denied per ambulatory visit (Brook and Williams, 1976).

Utilization, Costs, and Outcomes of Care

Both retrospective and prospective approaches have shown few systematic differences in use of resources and outcomes of care. Some studies suggest that family physicians are more efficient and effective providers of primary care whereas others show either no difference or an advantage of internists and pediatricians.

For example, there were few differences in a study of care in two clinics involving pediatricians. In one of the clinics internists and allied health workers formed teams with the pediatricians, providing a family orientation with both children and parents receiving care in the same clinic. In both clinics medical records were reviewed for (a) overall use of the center, (b) immunizations performed, (c) processes and outcomes in three conditions (dental caries, iron-deficiency anemia, and respiratory infections), (d) parental perceptions of the child's health status, (e) parental perceptions of the child's behavior, (f) parental view of the center's services, and (e) parental attitudes regarding the efficacy of medical care. There were few measurable differences between the two groups of children in utilization or outcome. Where differences existed, they tended to favor the family approach: older children stayed with the clinic rather than going elsewhere as they grew older, immunizations were somewhat more timely, and children appeared to make fewer visits over time to sources of care other than the center (San Agustin et al., 1981).

Another study involving pediatricians found differences. This study used a prospective design to assess the relative competence of pediatricians and general practitioners in managing febrile illnesses among 259 children under age ten, one group in an emergency room of a children's hospital and the other in a general hospital in Canada. Both groups of children were similar in their demographic characteristics, their presenting complaints, and the degree and duration of fever. A telephone interview was conducted within two weeks of the visit to determine outcomes based on duration of the acute illness and further physician contacts or hospitalizations. Although no significant differences in the measured outcomes of febrile illness were found, a trend favoring the pediatricians was discernible with two measures: unresolved symptoms at two weeks (8.3 versus 12.5 percent) and subsequent hospitalization (.8 versus 3.0 percent). The overall frequency of laboratory use and antibiotic prescriptions were the same for both types of physicians, but significant differences were found in the type of test: general practitioners ordered more x-rays and fewer microbiologic tests than the pediatricians (Leduc and Pless, 1982).

Case control techniques have found little systematic advantage of one type of physician over another in the care of children. In one such study, pediatricians and family physicians were compared with regard to their ability to recognize severe acute illness or avoid preventable complications. Children who had contacted a physician more than twenty-four hours before their emergency room visit were divided into two groups, one with acceptable outcomes and the other with potentially preventable complications or delayed diagnosis, treatment, or referral. Cases and controls were matched for age, socioeconomic status, and illness type. On follow-up interviews with patients, there was no evidence of better care by pediatricians than with family physicians, even taking into account other factors that might have influenced the results (Kramer et al., 1984).

Another case control study used medical records and telephone interviews to assess utilization and costs of care in a group of forty-five patients of family physicians matched with sixty-three patients attending medical and pediatric clinics over a thirty-three month period. Despite a 25 percent greater prevalence of significant chronic medical problems, family practice patients used specialist care less than half as much as other patients (.9 versus 1.8 visits per year), although they made an average of one more visit per year because of higher visit rates to their family physician (4.4 versus 2.5 visits per year). Costs for procedures did not differ (Farrell et al., 1982).

Studies of adult health care have also failed to show systematic differences between internists and generalists. In one study, charges for ambulatory and continuing patient care prescribed by residents in internal medicine were compared with those in family medicine for visits of patients with one of four diagnoses: congestive heart failure, diabetes, degenerative joint disease, or hypertension. Medical records of 4,991 encounters in internal medicine clinic and 700 encounters in the family medicine clinic were audited. The charges per encounter in the internal medicine clinic were greater than those in the family medicine clinic after differences in patient age, diagnosis, and severity of condition were accounted for. However, family practitioners scheduled follow-up visits more frequently with shorter durations between visits. Since the total duration of medical care for the selected problems (maximum of eighteen months) was shorter in family medicine for only two of the conditions and the number of visits for each condition within the total study period was greater, total charges over the period of the study were similar for the two types of residents even though the charges per visit were much greater for the patients of the internists (Bennett et al., 1983).

Another study compared the effects of physician specialty and board certification on costs and outcomes of health care for 213 patients with chronic lung disease followed for one year. Patients' pulmonary function, functional ability, number of medical conditions, and insurance status were predictors of outcome, but neither physician specialty nor board certification had a significant impact on outcomes or costs after the other factors were controlled for (Strauss et al., 1986).

Family physicians and internists do not appear to differ in their hospitalization practices, even controlling for several characteristics related to severity of illness. For example, one study of 523 hospitalized patients indicated that the length of stay and readmission rate in the intensive care unit and hospital, severity of illness, discharge diagnosis, proportion who died, time until death, consultation rate, and hospital charges did not differ significantly between patients cared for by the two different types of physicians (Hainer and Lawler, 1988).

Another study contrasted the inpatient treatment of diabetic ketoacidosis by internists and by family physicians in a teaching hospital. Laboratory use and length of stay were studied in twelve patients cared for by internists and in sixteen patients cared for by family physicians. Hospitalization was longer and the total number of laboratory and x-ray procedures per patient was greater in the internal medicine group. Serum glucose levels and urine spillage were comparable (Hamburger et al., 1982).

Several studies in England examined the outcomes of care in patients with diabetes. In one, there was no difference in blood glucose or glycosylated hemoglobin between office-based general practitioners and the hospital-based internists, but the patients followed by the general practitioners were more likely to have begun with less severe disease that was easier to control (Ruben et al., 1982). In another, no differences in the outcome of care were found for diabetic patients seen in hospital specialty clinics and for those seen in general practices that organized special clinics ("mini-clinics") for them (Singh et al., 1984). In a third study of patients randomized to general practice or to hospital-based internists' care, some differences were found. Patients cared for by the general practitioners had fewer blood glucose tests, higher glycosylated hemoglobins (indicating poorer control), more hospitalizations, and more deaths after a period of five years (Haynes and Harries, 1984).

Many of these studies are limited in study design, including a lack of power to detect true differences and difficulties in controlling for differences in case mix (Bowen, 1989). Nevertheless, the findings are strikingly inconsistent. This suggests that whatever differences exist among the different types of primary care physicians in profiles of care, patient satisfaction, technical quality, or utilization, the costs and outcomes of care may have more to do with factors *other* than the type of primary physician—family physician, internist, or pediatrician.

Chapter 12 indicates that the most important determinants of the quality of care are the length of postgraduate training, the extent of experience with the particular problem under consideration, and the nature of the organization in which the physician works (Palmer and Reilly, 1979). The diverse roles of the physician may heavily influence decision making, not always in consistent directions. Eisenberg (1986) characterized these roles as "self-fulfilling practitioner," "patient's agent," and "guarantor of social good." Sometimes one role dominates while at other times another prevails. Physicians may have needs for "self-fulfillment" because what they do may affect their income, certain ways of practice may be more personally satisfying to them, they have been trained in certain ways which are therefore more comfortable for them, the setting in which they practice has certain expectations that may be difficult to overcome, or their professional societies or colleagues recommend practicing in certain ways. Acting as the patient's agent may also produce differences in practice patterns because different physicians may interpret a patient's needs differently, patients may demand certain types of interventions, convenience to the patient may be considered a critical factor, or different types of patients are believed to need different types of interventions. As a "guarantor of social good" physicians may make decisions because of their beliefs about competing priorities. For example, they may be concerned about the benefit of an intervention when the required resources might be more beneficial to other patients. All of these considerations may contribute in a major way to differences in practice patterns among different types of physicians and to determining the best approaches to reducing the variability where it seems desirable to do so.

The Role of the "Team" in Primary Care

In some situations, largely in the developing nations, primary care providers are often not physicians. They may be nurses working in the community, pharmacists, or personnel trained specifically for a role that does not require a traditional medical education. The tasks that are required and the available resources determine the type of personnel involved (Kaprio, 1979). In industrialized nations, nonphysician personnel do not play a significant role in delivering primary care, at least as defined as the attainment of all the unique and essential features of primary care. However, they may play an important complementary role in improving certain aspects of primary care such as accessibility and comprehensiveness.

Few physicians work alone; most work with at least one nonphysician who interacts with patients in the office setting. Some of these staff members merely greet patients, make their appointments, or obtain and record administrative data. Others participate in the clinical encounter by carrying out the physician's instructions, such as taking x-rays or drawing blood for laboratory tests, administering medications or immunizations, or helping the physician perform some surgical procedure.

In some places, nonphysician staff members function more independently, even seeing patients themselves under the general guidance of the physician or following training to enable them to perform certain functions by themselves. In some cases, the nonphysician staff member performs functions that are unlikely to be performed well or with enthusiasm by physicians. For example, nurses may make home visits to help assess a patient's problem, to help a patient with a medical regimen, or to resolve some social problem that is interfering with the medical treatment.

Thus, nonphysician personnel perform three types of functions. One type performs a "supplementary" function, extending the efficiency of the physician by assuming some of the tasks, generally the technical ones, and usually under the direction of the physician. When the nonphysicians provide services that are often provided by physicians, they function as "substitutes." The third role is a "complementary" one, extending the effectiveness of physicians by doing things that physicians do not do at all, do poorly, or do reluctantly (Starfield, 1972).

In primary care, nurse practitioners or physician assistants are often the major providers of preventive care to children and adults, and they may undertake most of the responsibility for the care of uncomplicated acute and chronic illnesses. Because many physicians depend on these nonphysician practitioners (NPPs), when asked how many such individuals should be employed in practice, their response greatly exceeds the number currently employed (Weiner et al., 1986).

Most evaluations of the roles and effectiveness of NPPs were conducted during an era when a physician shortage was perceived and an effort was being made to train nonphysician personnel to compensate. These evaluations were consistent in supporting a role for such personnel in primary care, although the role has never been specified precisely enough to distinguish the supplementary, the substitute, and the complementary functions. The few evaluations that have been conducted demonstrated that trained NPPs could provide care for many acute and chronic conditions, as well as preventive care, at a level of quality that equaled or exceeded that provided by physicians (LeRoy, 1981; Record, 1981). NPPs perform some functions better than physicians: they identify more symptoms and signs in their patients and prescribe more nondrug therapies than physicians (Simborg et al., 1978). They also effectively help patients implement sustained and difficult regimens when therapeutic effects are often delayed (Starfield and Sharp, 1968).

Little is known about the extent of use of NPPs or the functions they perform within primary health systems in industrialized countries. One study in the United States showed great variability in their employment. For example, estimates of their supply for the country as a whole was 18.5 per 100,000 population (including all levels of care), 15.0 per 100,000 in nonmetropolitan areas, and 26.8 per 100,000 in three large HMOs (Weiner, 1990). However, the three HMOs differed markedly in the percentage of visits in which the patient was seen by an NPP. Table 9.2 shows

TABLE 9.2. Percentage of Visits in Which the Patient Was Seen by a Nonphysician Practitioner

Service	HMO #1	HMO #2	HMO #3
Adult medicine	15	6	47
General pediatrics	0	7	26
OB/GYN	21	41	56
Dermatology	0	0	60
General surgery	0	0	49
Orthopedic surgery	0	0	44

Source: Adapted from Weiner et al., 1986.

that patients enrolled in one of the HMOs were much more likely to see an NPP than in the other two HMOs.

It is apparent that primary care is, in most places, a physician-dominated effort that could not function without some teamwork involving nonphysician practitioners. Little is known, however, about the extent to which health care and health status are improved as a result of these team efforts, or under what conditions the team functions most effectively.

Implications: The Training of Primary Care Physicians

Primary care should be provided by physicians best trained and most skilled in its practice. The evidence summarized here and in Chapters 3 through 6 indicates that family physicians, general internists, and pediatricians are more effective and more efficient in providing primary care than specialists. Whether or not this superiority will continue depends on the ability of all primary care physicians to respond to the challenges posed by changes in illness patterns and to utilize the technology to prevent, cure, or ameliorate illness. With improvements in survival, the complexity of illness will increase. Illnesses with multisystem manifestations and multifactorial etiology will change the ways in which practitioners deal with patients' needs and how they relate to the community.

The primary care physicians of the future will become medical managers to a much greater extent than in the past. This will require skills in resource allocation, integration, and coordination, and in performance monitoring and quality assurance. Home care services will become more important, not for purposes of management of acute illness, but rather for the appropriate assessment of social factors complicating diagnosis and management and for the care of patients who are homebound as a result of functional disability. Advances in information technology can assist in diagnosis and management. They can also provide a better mechanism to obtain advice from consultants as well as to coordinate care. The skills to adapt and use these systems for primary care will have to be developed. Clinical epidemiology, decision making, economics, and training in the social and behavioral sciences related to health and health care will become part of the basic armamentarium in the education of the primary care physician. These challenges are readily recognized by leading educators of primary care physicians (Moore, 1990; Powe and Eisenberg, 1986). The success of primary care depends on society's adoption of these notions as central in its health policies for primary care training and delivery.

References

Aiken L, Lewis C, Craig J, Mendenhall R, Blendon R, and Rogers D. The contribution of specialists to the delivery of primary care: A new perspective. N Engl J Med 1979; 300:1363–70.

Bennett MD, Applegate WB, Chilton LA, Skipper BJ, and White RE. Comparison of family medicine and internal medicine: Charges for continuing ambulatory care. Medical Care 1983; 21(8):830–39.

Bertakis K and Robbins J. Gatekeeping in primary care: A comparison of internal medicine and family practice. J Fam Pract 1987; 24:305–09.

Bowman MA. The quality of care provided by family physicians. J Family Pract 1989; 28(3):346–55.

Brook R and Williams K. Evaluation of the New Mexico peer review system, 1971 to 1973. Med Care 1976; A4 (suppl):1–122.

Cherkin DC, Hart G, and Rosenblatt RA. Patient satisfaction with family physicians and general internists. Is there a difference? J Family Pract 1988; 26(5):543–51.

Cherkin DC, Rosenblatt RA, Hart LG, Schneeweiss R, and LoGerfo J. The use of medical resources by residency-trained family physicians and general internists. Med Care 1987; 25(6):455–69.

Cloher TP and Whelton PK. Physician approach to the recognition and initial management of hypertension. Arch Intern Med 1986; 146:529–33.

Eisenberg J. Doctors' Decisions and the Cost of Medical Care. Ann Arbor, Mich., Health Administration Press Perspectives, 1986.

Engel W, Freund D, Stein J, and Fletcher R. The treatment of patients with asthma by specialists and generalists. Med Care 1989; 27:306–14.

Farrell DL, Worth RM, and Mishina K. Utilization and cost effectiveness of a family practice center. J Family Practice 1982; 15(5):957–62.

Fishbane M and Starfield B. Child health care in the United States: A comparison of pediatricians and general practitioners. N Engl J Med 1981; 305:552–56.

Franks P and Dickinson JC. Comparisons of family physicians and internists: Process and outcome in adult patients at a community hospital. Med Care 1986; 24:941–48.

Fullard E, Fowler G, and Gray M. Promoting prevention in primary care: controlled trial of low technology, low cost approach. Br Med J 1987; 294:1080–82.

Hainer BJ and Lawler FH. Comparison of critical care provided by family physicians and general internists. JAMA 1988; 260(3):354–58.

Hamburger S, Barjenbruch P, and Soffer A. Treatment of diabetic ketoacidosis by internists and family physicians: A comparative study. J Fam Practice 1982; 14(4):719–22.

Hayes T and Harries J. Randomised controlled trial of routine hospital clinic care versus routine general practice care for Type II diabetics. Br Med J 1984; 289:728–30.

Hulka BS, Kupper LL, and Cassel JC. Physician management in primary care. Am J Public Health 1976; 66(12):1173–79.

Joffe A, Radius S, and Gall M. Health counseling for adolescents: What they want, what they get, and who gives it. Pediatrics 1988; 82 pt. 2:481–85.

Kahnemann D, Slovig P, and Tversky A. Judgement under Uncertainty: Heuristics and Biases. Cambridge, Cambridge University Press, 1982.

Kaprio L. Primary Health Care in Europe. World Health Organization, Copenhagen, 1979.

Kessler L, Amick B, and Thompson J. Factors influencing the diagnosis of mental disorder among primary care patients. Med Care 1985; 23:50–62.

Kohn R and White K. Health Care: An International Study. London, Oxford University Press, 1976.

Kramer MS, Arsenault L, and Pless IB. The use of preventable adverse outcomes to study the quality of child health care. Med Care 1984; 22(3):223–30.

Leduc DG and Pless IB. Pediatricians and general practitioners: A comparison of the management of children with febrile illness. Pediatrics 1982; 70(4):511–15.

LeRoy L. The Costs and Effectiveness of Nurse Practitioners. U.S. Congress, Office of Technology Assessment. Case Study #16. Washington, D.C., 1981.

Moore G. The case of the disappearing generalist: Does it need to be solved? Milbank Q 1992; 70(2):361–79.

Noren J, Frazier T, Altman I, and DeLozier J. Ambulatory medical care: A comparison of internists and family-general practitioners. N Engl J Med 1980; 302(1):11–16.

Palmer H and Reilly M. Individual and institutional variables which may serve as indicators of quality of medical care. Med Care 1979; 18:693–717.

Payne B, Lyons T, and Newhaus E. Relationships of physician characteristics to performance quality and improvement. Health Serv Res 1984; 19:307–32.

Powe N and Eisenberg J. Studying and teaching the gatekeeper. JGIM 1986; 1:197–98.

Rauber A and Maroncelli R. Prescribing practices and knowledge of tricyclic antidepressants among physicians caring for children. Pediatrics 1984; 73(1):107–09.

Record J. (ed.) Staffing Primary Care in 1990. New York, Springer Publishing Company, 1981.

Ruben L, Heller R, Jones R, Kelson M, Pearson T, and Shenouda F. Control of diabetics in general practice and hospital clinics. Practitioner 1982; 226:266–67.

San Agustin M, Sidel VW, Drosness DL, Kelman H, Levine H, and Stevens E. A controlled clinical trial of "family care" compared with "child only care" in the comprehensive primary care of children. Med Care 1981; 19(2):202–22.

Scherger JE, Gordon MJ, Phillips TJ, and LoGerfo JP. Comparison of diagnostic methods of family practice and internal medicine residents. J Fam Practice 1980; 10(1):95–101.

Simborg D, Starfield B, and Horn S. Physicians and non-physician health practitioners: The characteristics of their practices and their relationships. Am J Pub Health 1978; 68:44–48.

Singh B, Holland M, and Thorn P. Metabolic control of diabetes in general practice clinics: Comparison with a hospital clinic. Br Med J 1984; 289:726–28.

Smith DH and McWhinney IR. Comparison of the diagnostic methods of family physicians and internists. J Med Ed 1975; 50:264–70.

Sobal J, Muncie Jr HI, Valente CM, Levine DM, and DeForge BR. Self-reported referral patterns in practices of family/general practitioners, internists, and obstetricians/gynecologists. J Comm Health 1988; 13(3):171–83.

Starfield B. Higher education and the nation's health. Intl J Health Services 1972; 2:139–43.

Starfield B, Hoekelman RA, McCormick M, Mendenhall RD, Moynihan C, Benson P, and DeChant H. Styles of care provided to children in the United States: A comparison by physician specialty. J Fam Practice 1985; 21(2):133–38.

Starfield B and Sharp E. Ambulatory pediatric care: The role of the nurse. Medical Care 1968; 6:507–15.

Steinwachs D, Weiner J, Shapiro S, Batalden P, Coltin K, and Wasserman F. A comparison of the requirements for primary care physicians in HMOs with projections made by the GMENAC. N Engl J Med 1986; 314:217–22.

Strauss MJ, Conrad D, LoGerfo JP, Hudson LD, and Bergner M. Cost and outcome of care for patients with chronic obstructive lung disease. Med Care 1986; 24(10):915–24.

Thompson H and Osborne C. Office records in the evaluation of quality of care. Med Car 1976; 14:294–314.

Weiner J. HMOs and Managed Care: Implications for Rural Manpower planning. J Rural Health, 1991; 7(4 Suppl):373–87.

Weiner J, Steinwachs D, and Williamson J. Nurse practitioner and physician assistant practices in three HMOs: Implications for future U.S. health manpower needs. Am J Public Health 1986; 76:507–11.

10

Medical Records and Information Systems in Primary Care

The information you have is not what you want, the information you want is not what you need, and the information you need does not exist.

An Irish wit

To a large degree, medical care depends on a transfer of information. For this reason medical systems and professionals must maintain and provide information, both general and specific, so that other systems as well as other professionals who provide care or assess quality of care not only have access to information but can find it in a form applicable to the general practice of medical care and its associated aspects. In addition, information transfer is often required for medical–legal purposes, to document care provided.

A quite different type of information transfer involves the relationship between patient and physician. This chapter discusses the former aspects, and the following chapter considers the latter ones.

Medical records and information systems serve four functions. First, they are important as an aid to the memory of practitioners in caring for patients and also as an epidemiologic tool for planning care for populations. Second, medical records are important legal documents: what goes into records is considered to reflect the processes of care and therefore provide evidence when these processes are called into question. Third, medical records influence the processes of care; Chapter 6, in discussing coordination of care, documents how the contents of medical records affect what patients know about their care and how they respond to that knowledge. Fourth, medical records serve as a source of information about not only the quality of care but also how to improve it.

Since primary care is long-term rather than episodic as hospitalizations, the demands on the record differ from those in inpatient care. They must facilitate review of the patient over a period of time rather than at only one point. As yet, information science has not perfected techniques optimally suited to this challenge. Most uses of medical records still focus on review of particular visits or, at best, episodes of illness. These uses are further limited by their focus primarily on certain processes of care, most notably the processes of diagnosis and treatment within a visit or within an episode. As a result, records are less helpful in examining the practitioners' recognition of patients' needs, their understanding, acceptance, and participation in the processes of care, or the process of reassessment that spirals the cycle of care over time.

There is little doubt of the importance of records in clinical care. Records are often better than physicians' recollections for some types of information. For example, when patients were followed to determine their health status six months after a hospitalization, the accuracy of a prediction about their prognosis made at the time of discharge depended on the basis for the prediction. If the prediction was based on information from the medical record, it was often more accurate than when it came from recollections of the physician who had cared for the patient (Linn et al., 1974). Even though medical records have systematic deficits of information (Thompson and Osborne, 1976; Zuckerman et al., 1975), they do generally reflect many aspects of care. These include important problems that have been identified by both patients and physicians as requiring follow-up, medications (when the physiologic effects of medication are major), and tests with abnormal results (Starfield et al., 1979). The patient's chief complaint, at least as interpreted by the practitioner, is almost always recorded; information related to the history of the present illness is recorded about three-quarters of the time. Elements of the past medical history are recorded infrequently (Romm and Putnam, 1981) and certain types of information, such as social histories, are rarely present in the medical record (Chamberlin, 1971).

In its most complete form, the traditional medical record is a handwritten document of the patient's visit and contains a summary of the patient's "chief complaint," a section of variable length containing the "past medical history," a section containing the history related to the complaint, followed by a section labeled "physical examination" (which is often limited to describing the state of the part of the body most related to the chief complaint). There is also a section for "diagnoses" or "impressions," which can contain any number of items with no requirement that they be in any standard or codable form; yet another section is called a "plan" or "disposition." These practitioner notes are in one section of the record; other sections include laboratory reports and consultation notes. Records kept in this format are known as *source-oriented* because each item is inserted into a record according to its source. One modification of the source-oriented record is the *structured* medical record. Use of forms structured to require the recording of information in specific categories results in higher levels of performance as well as better recording (Duggan et al., 1990; Holmes et al., 1978; Cheney and Ramsdell, 1987). However, when the structured form is merely a list of items to be checked off, items may be recorded as done even if they were not (Duggan et al., 1990).

Another modification of source-oriented notes employs algorithms to guide and document care. In New York City such a system was implemented in several hospital-based primary care clinics. These guidelines improved documentation in medical records; they also improved adherence to guidelines in the management of several conditions for which the guidelines were developed, including asthma, otitis media, and gastroenteritis (Cook and Heidt, 1988).

Time-orientation is an alternative or, more usually, a supplementation to source-orientation. In a time-oriented chart, the information is entered by a particular item which is followed, over time, on a flow chart. This type of record is particularly suited to specialty care, in which a limited number of items are followed over time. The system used in the Rheumatology Clinic at Stanford University is the prototypical time-oriented record (Fries, 1974). These types of records, because of their organization and dependence on recording over time, often depend on computer technology.

Problem-orientation is a third format useful for primary care. A problem-oriented medical record (POMR) contains a problem list, usually in the front of the chart, consisting of a list of all currently active problems, updated at each visit. Resolved problems are crossed off. In some cases, the problem may be a symptom or sign that has not yet, and may never be, resolved into a standard diagnosis. In a complete POMR, each record contains a data base with a complete history, physical examination, and laboratory results at the initial visit. Each encounter is recorded in SOAP format. That is, for each problem listed on the problem list, there is recording of *s*ubjective data (what the patient says about the problem), *o*bjective data (what the practitioner finds on examination or testing), *a*ssessment (the practitioner's account of the status of the problem), and *p*lan (what is to be done about the problem). Although the full impact of the SOAP format has not been documented, the problem list improves the recognition and follow-up of problems from one primary care visit to the next particularly if different practitioners are seen (Simborg et al., 1977). The list of problems, as well as therapies and follow-up appointments, readily lends itself to computerization; these computerized lists provide the same benefits as do the handwritten lists (Johns et al., 1977).

During the 1980s, some medical facilities adopted computerization, but generally for fiscal and billing purposes rather than for clinical or research purposes. The best known and earliest of the clinical prototypes is the Computer-Stored Ambulatory Record (COSTAR), which has been used for many years in the Harvard Community Health Plan and is fully operational in ten of the twelve health centers in the Boston area. The system allows information transfer across the facilities, so that the records of individuals who visit another facility are always available to the physician seeing them (Schoenbaum, personal communication, August 1990). This record replaces all handwritten or typewritten medical records; there is no medical record except for information that cannot be computerized, such as electrocardiograms and other graphic material, and full hospital discharge summaries. The computer receives input when the practitioner records information on a self-encoding form specifically designed for the needs of each specialty in the health plan. There may be free-text entries, made either in writing or by dictation. Computer output is available in three forms: an encounter form containing data from individual encounters, a status report summarizing the current medical status of the patient, and flow charts showing the change in a given item over time. The record includes billing information, accounts receivable, electronic claims transfer, scheduling of personnel and resources, preventive services guidelines, and reminders to physicians of needed vaccines (Barton and Schoenbaum, 1990), as well as medical information. It provides summary counts, has a reminder system, and manipulates data for research purposes (Barnett, 1984). The system is now compatible with IBM computer systems. It is available for use in individual or group practices as well as for HMOs.

The TMR (The Medical Record) system was developed at Duke University. It includes an appointment system, diagnostic workups, treatments, problem lists, both subjective and objective findings, tests ordered, and accounts management. The data can be entered by physicians, nurses, secretaries, and technicians. Seven types of data are included: demographic information, medical history, problems, diagnoses, time-oriented data, studies and treatments, and encounter–admission linkages. The TMR is based on a set of modules which can be used either separately or together (Stead and Hammond, 1988).

A system developed in Indianapolis (the Regenstreif system) includes all inpatient data, outpatient data, emergency room visits, laboratory values, imaging reports, diagnoses, treatments, and pharmacy data. The system includes reminders for needed procedures and medications, and it flags items that may require special attention, such as potentially inappropriate treatments (McDonald et al., 1988).

The Summary Time Oriented Record (STOR) was developed at the University of California Hospitals in San Francisco. The system links the medical record to data from radiology, pathology, the clinical laboratory, microbiology laboratory, electrocardiogram facility, and the patient registration system (Whiting-O'Keefe et al., 1985).

Computerization of certain types of information facilitates some aspects of care even in the absence of a computerized medical record. For example, the use of microcomputer-generated reminders by primary care physicians increases the ordering of indicated mammograms (Chambers et al., 1989) as well as of other indicated preventive services such as tetanus immunization, Pap smears, breast examinations, and tests for occult stool blood (Turner et al., 1989).

Although computerization can facilitate the use of medical record information, primary care research using records long antedated computerization. The first systematic attempt to make physician records useful for research occurred in England by the Royal College of General Practitioners. Morbidity studies carried out in the mid-1950s by the College and the General Register Office of England and Wales showed that a substantial proportion of diagnoses in primary care could not be assigned disease codes and that almost half (45 percent) of the conditions seen in general practice could not be classified by the International Classification of Diseases (ICD) because they were symptom complexes rather than diseases. Because they could not be coded, it was impossible to conduct studies of their natural history and their impact of medical care. The college subsequently supported the development of a coding scheme useful for primary care and applied it to a method of indexing called the E-book (for Eimerl, its developer). The system consists of a series of small sheets, one for each patient, inserted into a looseleaf ledger divided into separate sections according to diagnosis. It includes information about the patient and type of encounter, and has several spaces that can be used to code items of special interest to the practitioner. The system can provide the number of episodes of illness and the number of visits per episode over a given time period. When combined with an age–sex register (a list of all patients in the practice, with their sex and age), E-book data was useful for epidemiologic research as well as for clinical investigations (Last, 1965). It did not, however, solve the problem of the inability to code symptoms.

Subsequent revisions of the ICD included codes for symptoms making it at least

TABLE 10.1.   Purposes of Minimal Data Set (Ambulatory Care)

To assist practitioner in caring for patients and managing their practice.
To facilitate self-evaluation.
To facilitate the collection of information on the natural history of health
problems.
To assist administrators in planning, allocating personnel, and monitoring costs.
To assist medical education by clarifying objectives of curricula.
To help local and regional planning efforts.
To permit uniform and interchangeable insurance claim forms.
To provide epidemiologists and health services researchers with sampling
frames.

Source: USDHEW, Vital and Health Statistics, Documents and Committee Reports. Ambulatory Medical Care Records: Uniform Minimum Basic Data Set. Series 4 no. 16 DHEW Pub. No. (HRA) 75-1453, August 1974.

theoretically possible to track illnesses from their initial presentation to their eventual diagnosis.

Despite this potential, the ICD has not been very useful in primary care because it is a compendium of all possible diagnoses, most of them uncommon and rarely, if ever, seen in primary care practice. ICD is a cumbersome instrument not readily useful to primary care practitioners, at least for purposes other than those of assigning a diagnosis for billing purposes. In 1979, the World Organization of National Colleges, Academies, and Academic Associations of General Practitioners/Family Physicians developed a classification specifically for use in primary care (WONCA, 1979). This classification, known as International Classification of Health Problems in Primary Care (ICHPPC-2), is compatible with the ICD but is much briefer and easier to use in clinical practice. Thus, the mechanism for using data on problems and disorders has been available since the 1980s.

Although many researchers recognized the need for a better information base, it was the involvement of the government in financing and providing care directly that brought it widespread attention. If public programs were to be accountable to the taxpayers, there had to be a method of generating and using information. In the late 1960s, the National Center for Health Statistics sponsored a series of conferences, which led to the publication of three sets of "minimal data": a set for long-term care, a set for inpatient care, and a set for ambulatory care. Table 10.1 presents the eight broad purposes of the patient's medical record, as identified in these conferences.

Subsequently, in 1975, the U.S. National Committee on Vital and Health Statistics, a group with the responsibility for recommending standard collection procedures for the country, proposed a minimum set of items to be collected for each patient encounter and presented three reasons for their recommendations:

1. Fewer and fewer providers of care practice as individuals, and hence there is an increasing need for records that are complete and comprehensible to colleagues.

2. An increasing proportion of ambulatory care costs are assumed by third-party payors who require a uniform set of information on which to base payments.

3. Interest in the quality and costs of care requires data that are comparable among providers.

TABLE 10.2.   Minimal Data Set for Ambulatory Care (NCHS)

A. Items that characterize the patient
   1. Patient identification
      a. Name
        Surname, first name, middle initial
      b. Identification number
        A Unique number that distinguishes the patient and his ambulatory medical care record
        from all others
   2. Residence
      Patient's usual residence, to consist of street name and number, apartment number (if any),
      city, state, zip code
   3. Date of birth
      Month, day, and year
   4. Sex
      Male or female
   5. Expected source of payment
      Government
      a. Workmen's compensation
      b. Medicare
      c. Medicaid
      d. Civilian Health and Medical Program of the Uniformed Services
      e. Other
      Nongovernment
      a. Blue Cross
      b. Blue Shield
      c. Insurance Company
      d. Prepaid group practice or health plan
      e. Medical foundation
      Self-pay
      No charge (free, charity, special research, teaching)
      Other
B. Items that characterize the provider
   1. Provider identification
      a. Name
        Surname, first name, middle initial
      b. Identification number
        A unique number that distinguishes the provider from all other providers
   2. Professional address
      Street address, office number (if any), city, state, and zip code
   3. Profession
      The profession in which the provider is currently engaged
      a. Physician
        Include specialty, if any as determined by membership in, or eligibility for, specialty board
      b. Dentist
        (include specialty)
      c. Nurse
      d. Other (specialty)
C. Items that characterize the patient–provider encounter
   1. Date of encounter
      Month, day, and year
   2. Place of encounter
      a. Private office
      b. Clinic or health center (any except hospital outpatient department)

(continued)

TABLE 10.2. *(Continued)*

---

    c. Hospital outpatient department
    d. Hospital emergency room
    e. Home
    f. Other (specify)
3. Reason for encounter
    The patient's principal problems, complaints, or symptoms on this encounter, in the patient's own words
4. Findings
    All history, physical examination, laboratory and other findings pertinent to the patient's reasons for visit or diagnoses, or both, and any other findings the provider deems important
5. Diagnosis and/or problem
    The provider's current assessment of the patient's reasons for the encounter and all conditions requiring treatment, with the principal diagnosis and/or problem listed first. Principal diagnoses and/or problem is defined as the health problem that is most significant in terms of the procedures carried out and the care provided at this encounter
6. Services and procedures
    All diagnostic, therapeutic, and preventive services and procedures (including history taking) performed during the encounter and those scheduled to be performed before the next encounter
7. Itemized charges
    All charges to be made by the provider for services and procedures performed during the encounter or to be performed by him or his associates before the next encounter
8. Disposition (one or more)
    The provider's statement of the next step(s) in the care of the patient
    a. No follow up planned
    b. Return, time specified
    c. Return, P.R.N.
    d. Telephone follow up
    e. Referred to other provider
    f. Returned to referring provider
    g. Admit to hospital
    h. Others

Source: USDHEW, Vital and Health Statistics, Documents and Committee Reports. Ambulatory Medical Care Records: Uniform Minimum Basic Data Set. Series 4 No. 16 DHEW Pub. No. (HRA) 75–1453, August 1974.

The details of the minimal data set recommended by the National Committee are in Table 10.2.

Another contribution of the National Center for Health Statistics was the initiation of a major undertaking in a series of continuing national assessments of health and health services. This was the National Ambulatory Medical Care Survey (NAMCS), a nationwide effort that requests office-based physicians (selected to represent the various specialties) to complete one-page forms for a sample of patients seen during one week. The physician is requested to keep a log of all patients, and is instructed on how to choose the sample. Although some specific items have been added in different years, the basic set of information consists of fifteen items:

- Date of visit
- Patient's date of birth
- Sex

- Color or race
- Ethnicity
- Expected source of payment for the visit
- A query as to whether the patient was referred for this visit
- The patient's complaint(s), symptom(s), other reason(s) for visit
- Other diagnostic services this visit
- Physician's diagnosis
- A query as to whether the physician had seen the patient prior to this visit
- Nonmedication therapy
- Medication therapy
- Disposition
- Duration of visit (time that the patient spent with the physician)

Data from this survey are published periodically and provide a view of encounters in office-based practice throughout the country. Because it provides information on encounters, it is not suitable as a source of information about the natural history of patient's complaints nor about the evolution of illness or combinations of illnesses in individuals over time. It is a good mechanism for describing practices and the burden particular conditions place on them, but it is not a mechanism for obtaining information about the effectiveness of medical care or about the patterns of illness in the population. Although the minimum ambulatory care data set and the NAMCS are designed for all of ambulatory care, including specialty care, they contain one element of information especially important in primary care: the patient's reason for the visit. Parallel efforts at the National Center for Health Statistics produced a framework to code these reasons (Schneider et al., 1979).

The pioneering work of the National Center stimulated an international collaborative project to devise and test a system of nomenclature for problems in primary care (Lamberts and Wood, 1987). The basis for the coding scheme, known as the ICPC, is a biaxial alpha-numeric structure in which there are "components" and "chapters." Each reason for visit is categorized as one of seven types:

- Symptom or complaint
- Diagnostic, screening, or preventive procedure
- Treatments, procedures, or medications
- Receipt of test results
- Administrative
- For a specific diagnosis
- All others

Each reason is also assigned to one of seventeen chapters indicated by letter, fifteen of which are body systems, one for problems that cross systems or are general in nature, and one for social problems. Within the chapters are specific codes, organized in a similar way across the chapters according to the appropriate component. For example, the first twenty-nine codes in each chapter refer to specific symptoms or complaints; codes 30–49 refer to specific diagnostic procedures; codes 50–59 are therapeutic modalities; codes 60 and 61 are, respectively, results of tests or procedures and results from other providers; code 62 is an administrative procedure;

codes 63–69 are various other reasons (including the institution of a referral or an encounter initiated by someone other than the patient); and codes 70–99 are the most common specific diagnoses within the chapter. The coding system for the symptoms/complaints and diagnosis components is compatible with the International Classification of Disease, the International Classification of Health Problems in Primary Care (ICHPPC-2), and the Classification of Diseases, Problems, and Procedures 1984 of the British Royal College of General Practitioners. The availability of uniform systems to code presenting problems and reasons for visits will greatly facilitate research in primary care and lead to greater understanding of the distribution of health problems and the impact of various methods of managing them. It will also focus attention on a most critical but relatively neglected aspect of care: recognition of the patients' problems by practitioners.

References

Barnett GO. The application of computer-based medical record systems in ambulatory practice. N Engl J Med 1984; 310:1643–50.

Barton M and Schoenbaum S. Improving influenza vaccination performance in an HMO setting: the use of computer-generated reminders and peer comparison feedback. Am J Public Health 1990; 80:534–36.

Chamberlin R. Social data in evaluation of the pediatric patient: deficit in outpatient records. J Pediatr 1971; 78:111–16.

Chambers C, Balaban D, Carison B, Ungemack J, and Grasberger D. Micro-computer generated reminders: improving the compliance of primary care physicians with mammography screening guidelines. J Fam Practice 1989; 29:273–80.

Cheney C and Ramsdell J. Effect of medical records' checklists on implementation of periodic health measures. Am J Med 1987; 83:129–36.

Cook CD and Heidt J. Assuring Quality Out-Patient Care for Children: Guidelines and a Management System. New York, Oxford University Press, 1988.

Duggan A, Starfield B, and DeAngelis C. Structured encounter form: the impact on provider performance and recording of well-child care. Pediatrics 1990; 85:104–13.

Fries J. Alternatives in medical record formats. Medical Care 1974; 12:871–81.

Holmes C, Kane R, Ford M, and Fowler J. Toward the measurement of primary care. Milbank Mem Fund Q 1978; 56:231–52.

Johns C, Simborg D, Blum B, and Starfield B. A minirecord; an aid to continuity of care. Johns Hopkins Med J 1977; 140:277–84.

Lamberts H and Wood M (eds), ICPC. International Classification of Primary Care. World Health Organization of National Colleges, Academics, and Academic Associations of General Practitioners/Family Physicians. Oxford, Oxford University Press, 1987).

Last J. Primary medical care. Record keeping. Milbank Mem Fund Q 1965; 43 (Part 2):266–76.

Linn B, Linn M, Greenwald S, and Gurel L. Validity of impairment ratings made from medical records and from personal knowledge. Medical Care 1974; 12:363–68.

McDonald C, Blevins L, Tierney W, and Martin D. The Regenstrief medical records. MD Comput 1988; 5:34–47.

Romm F and Putnam S. The validity of the medical record. Medical Care 1981; 19:310–15.

Schneider D. An ambulatory care classification system: design, development and evaluation. Health Services Research 1979; 14:77–87.

Schoenbaum, personal communication, August 1990.

Simborg D, Starfield B, Horn S, and Yourtee S. Information factors affecting problem follow-up in ambulatory care. Medical Care 1976; 14:625–36.

Starfield B, Steinwachs D, Morris I, Bause G, Siebert S, and Weston C. Concordance between medical records and observations regarding information on coordination of care. Medical Care 1979; 17:758–66.

Stead W and Hammond W. Computer-based medical records. The centerpiece of TMR. MD Comput 1988; 5:48–62.

Thompson H and Osborne C. Office records in the evaluation of quality of care. Medical Care 1976; 14:294–314.

Turner B, Day S, and Borenstein B. A controlled trial to improve delivery of preventive care: Physician or patient reminders? J Gen Int Med 1989; 4:403–9.

Whiting-O'Keefe Q, Simborg D, Epstein W, and Warger A. A computerized summary medical record system can provide more information than the standard medical record. JAMA 1985; 254:1185–92.

WONCA (World Organization of National Colleges, Academies, and Academic Associations of General Practitioners/Family Physicians. ICHPPC-2 (International Classification of Health Problems in Primary Care). 3rd ed. Oxford, Oxford University Press, 1983.

Zuckerman A, Starfield B, Hochreiter C, and Kovasznay B. Validating the content of pediatric outpatient medical records by means of tape-recording doctor-patient encounters. Pediatrics 1975; 56:407–11.

11

Physician–Patient Interactions in Primary Care

Dr. K practices medicine on the lower east side of Manhattan. He says, "Many times patients will come in and given all sorts of exotic stories until they are brave enough to tell you the real reason they are here. Yesterday a patient came in and told me about aches and pains and headaches and problems in all parts of her body and she was sure it wasn't the flu and that there was something wrong with her and she didn't know what it was. I examined her but I couldn't find anything abnormal. So I kept encouraging her to just talk to me about what she was feeling. I noted an anxious look on her face and I watched her. Finally she broke down and said "Listen, I've been taking drugs." Then it became clear that the problems she was having were withdrawal symptoms. I don't think that if I had asked her a lot of questions, or even asked her about whether she was taking drugs, she would have told me. She had to work up the courage to do it and that took a little time and my encouragement. The thing is, if you can break the barriers down, you can find out why they really came to see you and solve the problem without a lot of visits, tests, and drugs. I could have said "You've got a lot of aches and pains, so I'll take a few tests and give you some aspirin to take home and come back in two days and let's see what happens." If you just listen and watch patients, you learn from them. When you see an anxious look, you know that the patient wants to tell you something, and you just have to let it come out when it's ready to come out. A doctor treats patients, not toilets, electrical appliances or air conditioners. A human being is complex, not a machine, so you really have to try to understand things. Anybody can order a test or write a prescription. The whole thing about practicing medicine is to find out why."

Information transfer within the context of the physician–patient relationship assumes a quite different form from that described in the previous chapter, which dealt with institutional and professional information transfer. This chapter is devoted to a description of how modern approaches to patient–physician information transfer in primary care greatly improve effectiveness of medical care with resultant benefit to both physician and patient.

Medical care begins with a transfer of information. Patients convey information so that health practitioners can assess the nature, duration, and context of health problems. Practitioners convey information to enable patients to understand their health problems, their management, and their prognosis.

People are often frightened to tell the doctors things. They might get angry, or say you are imagining it, or think it is trivial. A lot of people are afraid of doctors. Maybe the doctor will classify them as a "mental patient." Because they are afraid, they learn to hide things from doctors, assuming that the doctors will find out anyway because of their training. I think doctors should act as human beings, not as doctors. Doctors should know that each patient is a living person, with feelings. Even if the doctor has treated the same medical problem over and over again, each patient is different.   Mr. M, a fifty-year-old contractor

The model of the health services system presented in Chapter 2 provides the underpinning for much of the approach to measuring the various aspects of primary care. This section returns to that model, particularly the part that involves both providers and patients in an interaction contributing to the processes of care. When patients seek care, they present themselves and their problems, concerns, and needs by using the system (utilization); the practitioner bears responsibility for recognizing the patient's problems, needs, or concerns (problem recognition) and for formulating those needs into a diagnosis for which appropriate management can be suggested. Patients participate through their understanding of the suggestions and recommendations and for accepting them as appropriate. They also participate in the suggested regimen by either carrying it out and returning at the recommended time or seeking alternatives when the recommendations do not appear appropriate or are inadequate. Practitioners then reassess the situation by determining whether the problem is resolved, the concerns ameliorated, or the needs adequately met.

The interaction between the processes contributed by professionals and those contributed by patients takes the form of communication occurring during the medical encounter. The physician–patient interaction is important at all levels of medical care, but especially in primary care. The long-term relationship that characterizes primary care will be difficult to sustain if either party is uncomfortable with their encounters. Since many problems that patients bring to primary care physicians are of uncertain cause or prognosis, the relationship must be strong enough to tolerate ambiguity, at least at some stages in the development of problems. This strength is built, in part, on a free flow of relevant information by both parties and on the rapport that provides the context for the relationship. Much communication serves to convey specific items needed to deal directly with the problem, concern, or need. Conversation is the most common mode of communication. The practitioner asks questions to elucidate the problem, and informs the patient about the diagnosis, the plan of intervention directed at resolving or ameliorating the problem, and the plan for reassessment. Patients provide information about their problems, concerns, and needs, and later query the physician if they are unclear about aspects of the diagnosis, therapy, or follow-up plan.

This is the ideal situation. In practice, short-cuts may occur or patients may not be permitted to say as much about their problem as they would like to, either because the practitioner directs the flow of information by asking only certain questions or because the patient is not allowed enough time. Or the practitioner may provide too little information, leaving the patient unable to understand the diagnosis, the plan of intervention, or the plan for reassessment.

In traditional theories of physician–patient interaction, the practitioner's role received a disproportionate share of attention. Social theorists assumed that the physician had a commanding role because of the functions ascribed to the physician–patient interaction. Some theorists considered the physician to be an agent of social control who legitimized a "social deviance," such as absence from work because of illness. According to these traditional theories, the role of the patient is relatively passive and that of the professional is dominant. Others believed that the physician's economic interests dictate keeping patients satisfied to prevent their seeking care from other sources. This requires the maintenance of a "professional" demeanor involving specialized knowledge and an arcane language. In these theories, the patient might not be passive but any conflict between the patient and physician is minimized to enhance the physician's stature (Freidson, 1972).

The assumption by the physician of a dominant role interferes with the appreciation of other influences on patients and with the fact that patients do not necessarily regard physicians' advice as the most salient. Moreover, physicians and patients often disagree on the nature of the patient's problem; in general, physicians and patients are in agreement only about half of the time (Starfield et al., 1979b; Starfield et al., 1981; Connelly et al., 1989). Only about half of all patients report playing an active role in the interaction with their physician; the other half believe their role to be passive (Brody et al., 1989a).

> Mr. R is a forty-two-year-old engineer who experienced a sharp pain in his chest when lifting a 20-pound load. Because he has a family history of heart disease, he sought care from a cardiologist. The following dialogue ensued:
> *Physician:* "One of the advantages of being a cardiologist is that I can allay your fears. I've looked at your tests and I do not believe that you have had a significant coronary event.
> *Mr. R:* Thank you doctor. But you have not allayed my fears. I am an educated man and when I hear you use the words "believe," "significant," and "event" in one sentence, I worry about what you are telling me. I am not at all reassured.
> Charge for consultation: $400
> Charge for laboratory tests: $375

Some conceptualizations of the physician–patient interaction assign a larger role to the patient. Szasz and Hollander, for example, describe three models for physician–patient interaction. In the first, the activity–passivity model, the physician makes the decisions. This occurs in situations in which the patient is anesthetized or unconscious. In the second, the guidance–cooperation model, the physician provides the instructions and the patient carries them out ("cooperates"). In the third, the mutual participation model, the physician assists patients in helping themselves; patients participate in decision-making. In this third model, the physician does not presume to know what is best for the patient; patients are aware of constraints in their own life circumstances that make certain methods of intervention more or less feasible. What is best for the patient emerges from the interaction between patient and physician in which both contribute their own unique information and then negotiate the most appropriate approach to management (Szasz and Hollander, 1956).

The rise of the consumer movement and increasing sophistication of the population inevitably narrowed the gap between professional competence and consumer knowledge (Haug and Levin, 1981), leading to different types of exploration of the dynamics of the interaction.

Depending on the degree of practitioner control over the interaction versus that of the patient, four types of interaction are now recognized. When both the patient and the physician have high control, the interaction is characterized as "mutuality." When the patient has high control but the physician has low control, the interaction is "consumerist." Low patient and high practitioner control produces a "paternalistic" interaction, whereas low control by both participants produces a "default" interaction (Stewart and Roter, 1989).

Despite its importance, the verbal content of the interaction alone cannot sustain the practitioner–patient relationship. Both practitioners and patients respond to the context of the interaction, how each feels about the other, and how those feelings are conveyed. This context can be assessed by analyzing the voice tones rather than the context of the interaction, or by observing physical postures and mannerisms. Voice tone and affect have more influence on what patients carry away from their encounter than does the content of the interaction, even when specific items of information are concerned.

The recent literature on the dynamics of the medical encounter focuses on both parties in the encounter by examining the nature of both the verbal communication and nonverbal behaviors. Although communication between professional and patient is not limited to primary care, most knowledge about the subject derives from studies in primary care facilities. It is possible to code the content of the interaction as well as the tone of the vocal communication. Several methods of coding the components of the interaction are available (Roter and Hall, 1989), the two main categories of components being verbal information transfer between patients and physicians and socioemotional behaviors. Information transfer involves question-asking, giving information relevant to the illness or health problem and its management, as well as concordance on problem identification and treatment. Techniques for ascertaining and evaluating information transfer usually involve audiotaping the encounter, although direct observation and listening by a neutral observer has also been effective. The content of the verbal interaction is then analyzed. Socioemotional components involve explicit communication of feelings; feelings conveyed as nervousness or anxiety; warmth or empathy; or expressions of satisfaction and confidence. The socioemotional content of the interaction is also assessed from audiotapes, but here the voices are passed through an electronic filter that eliminates the highest voice frequencies and makes the verbal content unintelligible while preserving rhythm, tempo, and other nonverbal qualities (Hall et al., 1987).

The characteristics of the physician–patient interaction contribute substantially to what patients think and do, as reflected in their satisfaction, recall of information pertinent to their problem, understanding of that information, and compliance with advice.

Studies of these phenomena have shown that patient satisfaction with care is most related to the amount of information given by practitioners, especially of general rather than technical information (Roter and Hall, 1989). Examples of general information include explanation about the health problem and its treatment, stress counseling, and discussion of patient's ideas about the problem and its management. Technical information includes such items as the results of examinations or tests (Brody et al., 1989b). Patient satisfaction is also improved by more overall communication, especially by social conversation, partnership-building conversation, positive feelings, and positive talk, but it is reduced by conveying negative feelings or information.

Although physicians are taught to ask questions, most patients do not ask many questions even though they may want information. The challenge for the physician is to establish a dialogue in which patients' concerns emerge in a context other than direct questioning. Patients are more satisfied, with the technical as well as the interpersonal aspects of care, when physicians involve them in decision-making by asking whether the patient agrees with the decision about the problem, its causes, and its management. Physician-initiated interest in the patient's views has a greater impact on satisfaction than does the patient's perceived ability to ask questions or the patient's report of self-initiated comments. In fact, the patient's ability to ask questions or to initiate comments improves satisfaction only with the technical aspects of care and has no effect on the interpersonal aspects (Lerman et al., 1990). Thus, when physicians solicit the involvement of patients, the latter are more satisfied with the care, a phenomenon that supports the newer "sharing" theories of the physician–patient interaction rather than the older authoritarian models.

> I went to see the dermatologist and the nurse showed me inside the examining room and told me to take off my clothes and lie down on the table. The doctor came in and examined me. It was like being in a pathology lab. He was dictating a very beautiful description of my rash into a microphone in the ceiling. After he looked me all over, the first thing he said was "Well, I'm 99 percent sure that it's not a fungus but I'm going to take a test to be 100 percent sure." What he didn't tell me at the time was that the test would cost $20. He might have said "Would you like me to take the test now or if it doesn't go away in a few days to take it then?" but his nurse came in and scraped me with a little knife, and then the doctor told me to turn over. There I was naked on my stomach and he said "I'm going to give you a cortisone shot. I think that will probably take care of it or at least I hope it will take care of it." And then the nurse stuck me with a needle. Now I don't like to take a lot of medication but I didn't have much time to think about anything and when you are lying there with your behind sticking up, you're sort of vulnerable and so before I had a chance to decide anything, I had the shot. Then he handed me a bottle of pills and when I asked what they were he said that they were cortisone pills and he wrote out a prescription and told me these were for more pills. I then said "Do I have to take these pills? Maybe the shot will do it?" "Well," he said, "you don't have to take the pills," and he took them from me and threw them in the basket. So I said, "I'll just take the prescription and if the rash doesn't go away I'll get the prescription filled." "No," he says, "you know best," and he tore up the prescription.   Mr. C, a thirty-year-old accountant

Patients are more likely to accept advice and instructions when the practitioner provides more information, asks fewer questions overall (but more questions about compliance), and provides more positive and less negative feedback. In addition, patients have a better understanding of advice particularly when the practitioner makes efforts to build a partnership. Useful strategies for improving patient's understanding of advice include the following (Roter and Hall, 1989):

* Presenting information early in the visit.
* Taking care to be specific rather than general in instructions.
* Organizing information into clear blocks.
* Repeating the most important information.
* Asking patients for feedback as to their understanding.
* Summarizing information at the close of the visit.

The characteristics of the interaction between physicians and patients also influence the outcomes of medical care. When both agree on the nature of the problem, the outcome is more likely to be reported as improved, by both patient and physician at the time of a follow-up visit (Starfield et al., 1979; Starfield et al., 1981; Bass et al., 1986; Headache Study, 1986). In addition, when patients believe they were active participants in the medical encounter, they report less discomfort, greater alleviation of their symptoms, and more improvement in their general condition on follow-up a week after their visit. They also report less concern with their illness and a greater sense of control over it (Brody et al., 1989a; Wasserman et al., 1984). Objective signs of better disease control are also evident when patients actively participate in the physician–patient interaction. For example, greater patient involvement results in better control of their disease and improved quality of life (Greenfield et al., 1985, 1988; Roter, 1977; Kaplan et al., 1989; Orth et al., 1987).

Although the medical encounter has long been considered an art rather than a science, new avenues of research demonstrate how old techniques are amenable to study and change. There is a "technology" that can be taught. Talking with patients may be an ancient art, but there is also a modern science based on empirical evidence. Physicians trained in skills which make patients more active participants in the medical encounter obtain better diagnostic information from them (Maguire et al., 1986). This training also improves the quality of the physician's explanation to patients and enhances the usefulness of what the patient tells the physician (Putnam et al., 1988). Checklists to evaluate medical interviewing skills at the undergraduate level are available (Kraan et al., 1989). There are models for teaching doctor–patient communication during residency (Schofield and Arntson, 1989). Given such training, primary care residents are better able to detect psychiatric illness among their patients (Goldberg et al., 1980). If skills in communication were not part of medical training, they can be learned afterwards (Kurtz, 1989).

The adequacy of the patient–physician interaction includes more than the mere transfer of information. What patients know about their problems and their treatments is determined by more than what practitioners tell them. Illness has meaning to patients far beyond its physical manifestations because every illness produces some disturbance in the person's life. Cure removes the physical abnormality of an illness, but healing requires restoration of the relationships disturbed by it. If a practitioner does not attend to the nonphysical correlates of the illness by failing to explore their meaning to the patient, the outcome may be unsatisfactory (McWhinney 1989). Patients come with preconceptions that influence the way they interpret both their illness and the physician's advice. When practitioners explore with patients what they know and what they think about their illness, its effects, and its management, these preconceptions can be revealed. Then the physician can furnish information in a context meaningful to the patient. For this reason, straightforwardness in providing factual advice is not generally the best approach, although it may seem efficient (Maynard, 1990). The way information is conveyed to patients and their involvement as active participants in the dialogue have a powerful impact on what they think, how they feel, and how they respond to medical advice.

Since health is heavily influenced by the social context in which people live and work, social factors might be expected to be an important component of physician–patient interactions. Many studies have shown that this is not the case. The training of physicians generally focuses on biological determinants of illness rather than social determinants, and there is no accepted method of categorizing social factors as there is for biological factors. Waitzkin's philosophical and empirical analysis of the nature of the interaction between physicians and patients (Waitzkin, 1991) shows how ideology determines the content of the discourse and how it leads physicians to avoid the social issues that cause illnesses and influence their course and response to treatment. For example, a well-accepted goal of medical care is to return individuals to their jobs. Since patients may have different value systems and not share these goals, open discussion of strategies to reduce illness, disability, and discomfort can produce disagreement and conflict. Furthermore, the solution to many social problems lies in collective action leading to change in the social conditions that predispose to illness. In contrast, medical care generally approaches these social problems by reinforcing individual accommodation to social conditions rather than by encouraging efforts to change the social conditions. As a result, the underlying tension between doctors and patients remains unresolved. Waitzkin's short-term approach to changing this characteristic of medical discourse lies in a conscious attempt by physicians to recognize the existence of this underlying tension, to understand its basis, and to avoid technical solutions such as medications to reduce work-related stress and its physiological effects. Rather than "over-medicalizing" health problems, physicians could help patients to understand the genesis of their problems and encourage their involvement in collective rather than individual activities to deal with their social problems. In the long run, more fundamental restructuring of health systems to bring them into closer relationship with other social systems is needed to make medical care more effective in preventing and healing illness.

References

Bass M, Buck C, Turner L, Dickie G, Pratt G, and Robinson, H. The physician's actions and the outcomes of illness in family practice. J Fam Pract 1986; 23:43–47.
Brody DS, Miller SM, Lerman CE, Smith DG, and Caputo GC. Patient perception of

involvement in medical care: relationship to illness attitudes and outcomes. J Gen Intern Med 1989a; 4:506–11.

Brody DS, Miller SM, Lerman CE, Smith DG, Lazaro CG, and Blum MJ. The relationship between patients' satisfaction with their physicians and perceptions about interventions they desired and received. Medical Care 1989; 27(11):1027–35.

Connelly J, Philbrick J, Smith G, Kaiser D, and Wymer A. Health perceptions of primary care patients and the influence on health care utilization. Medical Care 1989; 27(3 Suppl):S99–S109.

Freidson E. Client control and medical practice. In: Jaco EG. Patients, Physicians and Illness. A Source Book in Behavioral Science and Health. 2nd ed. London, Collier MacMillan Publishers, 1972.

Goldberg D, Steele J, Smith C, and Spivey L. Training family doctors to recognize psychiatric illness with increased accuracy. Lancet 1980; 2:521–23.

Greenfield S, Kaplan S, and Ware JE. Expanding patient involvement in care. Ann Intern Med 1985; 102:520–28.

Greenfield S, Kaplan SH, Ware JE Jr, Yano EM, and Frank H. Patients' participation in medical care. J Gen Intern Med 1988; 3:448–57.

Hall J, Roter D, and Katz N. Task versus socioemotional behaviors in physicians. Medical Care 1987; 25:399–412.

Haug M and Levin B. Practitioner or patient—who's in charge? J Health Soc Behav 1981; 22:212–29.

Headache Study Group of the University of Western Ontario. Predictors of outcome in headache patients presenting to family physicians—a one year prospective study. Health Journal 1986; 26:285–94.

Kaplan S, Greenfield S, and Ware JE Jr. Impact of the doctor–patient relationship on the outcomes of chronic disease. In: Stewart M and Roter D (eds). Communicating with Medical Patients. Newbury Park, Calif., Sage Publications, Interpersonal Communications 9, 1989.

Kraan HF, Crijnen A, Zuidweg J, der Vleuten C, Imbos T. Evaluating undergraduate training—a checklist for medical interviewing skills. In: Stewart M and Roter D (eds). Communicating with Medical Patients. Newbury Park, Calif., Sage Publications, 1989.

Kurtz SM. Curriculum structuring to enhance communication skill development. In: Stewart M and Roter D (eds). Communicating with Medical Patients. Newbury Park, Calif., Sage Publications, 1989.

Lerman C, Brody D, Caputo C, Smith D, Lazaro C, and Wolfson H. Patients' perceived involvement in care scale: relationship to attitudes about illness and medical care. J Gen Intern Med 1990; 5:29–33.

Maguire P, Fairbairn S, and Fletcher C. Consultation skills of young doctors: I-Benefits of feedback training in interviewing as students persist. Br Med J 1986; 292:1573–76.

Maynard D. Bearing bad news. Medical Encounter Newsletter on the Medical Interview and Related Skills. Summer 1990; 7:2–3.

McWhinney I. The need for a transformed clinical method. In: Stewart M and Roter D (eds). Communicating with Medical Patients. Newbury Park, Calif., Sage Publications, 1989.

Orth J, Stiles W, Scherwitz L, Hennrikus D, and Vallbona C. Patient exposition and provider explanation in routine interviews and hypertensive patients' blood pressure control. Health Psychology 1987; 6:29–42.

Putnam S, Stiles W, Jacob M, and James S. Teaching the medical interview: an intervention study. J Gen Int Med 1988; 3:38–47.

Roter DL. Patient participation in the patient provider interaction: the effects of patient question asking on the quality of interaction, satisfaction and compliance. Health Educ Monographs 1977; 5:281–315.

Roter DL and Hall JA. Studies of doctor–patient interaction. Ann Rev Public Health 1989; 10:163–80.

Sackett D, Haynes RB, and Tugwell P. Clinical Epidemiology. A Basic Science for Clinical Medicine. Boston, Little, Brown & Company, 1985.

Schofield T and Arntson P. A model for teaching doctor–patient communication during residency. In: Stewart M and Roter D (eds). Communicating with Medical Patients. Newbury Park, Calif., Sage Publications, 1989.

Starfield B, Steinwachs D, Morris I, Bause G, Siebert S, and Weston C. Patient–provider agreement about problems. Influence on outcome of care. JAMA 1979b; 242:344–46.

Starfield B, Wray C, Hess K, Gross R, Birk P, and D'Lugoff B. The influence of patient–practitioner agreement on outcome of care. Am J Public Health 1981; 71:127–32.

Stewart M and Roter D. Communicating with Medical Patients. Newbury Park, Calif., Sage Publications, 1989.

Szasz T and Hollander M. A contribution to the philosophy of medicine. Arch Int Med 1956; 97:585–92.

Waitzkin H. The Politics of Medical Encounters: How Doctors and Patients Deal with Social Problems. New Haven, Yale University Press, 1991.

Wasserman R, Inui T, Barriatua B, Carter W, and Lippincott P. Pediatric clinicians' support for parents makes a difference: an outcome-based analysis of clinician–parent interaction. Pediatrics 1984; 74:1047–53.

12

Quality Assessment and Quality Improvement

Social systems that are supported by and entrusted with public funds must be accountable for their performance. In the health system this accountability takes the form of quality assurance. This chapter describes methods of assessing the quality of care and their application to ensuring quality in the primary care setting.

Early approaches to assessing the quality of care depended heavily on medical records; they also focused primarily on inpatient care, probably because the records were more accessible and more standard in format than those in ambulatory care. The techniques used by early investigators, however, are equally applicable to primary care. One technique uses "profiles of care," in which information on a variety of characteristics such as diagnosis, operations, laboratory tests, and consultations are abstracted from medical records and used to make comparisons among hospitals in a given area (Eisele, 1956). Another technique, the medical audit, relies on the a priori development of criteria for good care. Lembcke (1956), for example, used the medical literature to define "good care" for selected diagnoses and developed lists of criteria for judging care of patients with information in the medical record. He proposed that every hospital maintain an internal audit based on criteria from the literature. Later the Joint Commission on the Accreditation of Hospitals required that, as a condition for accreditation, hospitals have committees to conduct internal audits, but did not specify what had to be audited or require that deficiencies be corrected.

In the 1960s, attention shifted to quality assessment in ambulatory care because the federal government began to assume responsibility for payment for services on a large scale. Medical audits were mandated for government-funded direct service programs, and demonstration programs were supported to develop the procedures (Morehead and Donaldson, 1974). By and large, medical audit procedures were employed, although the choice of diagnoses differed from those used for inpatient care; common targets for medical audits included well-infant and well-child care; recognition and management of iron-deficiency anemia and asthma in children, hypertension, diabetes, and peptic ulcer in adults; high-risk pregnancies in obstetrics; and pelvic inflammatory disease or menopausal symptoms in gynecology. Aspects that were scrutinized included justifiability of the diagnosis, components of the physical examination, indicated laboratory or x-ray studies, acceptability of therapy, and follow-up procedures and visits.

Payne and Lyons (1984) used a similar technique in studies of the quality of care in five different types of primary care facilities, which included gynecology, family medicine, internal medicine, and pediatrics. The five sites were a solo practice, two university teaching clinics, a fee-for-service multispecialty group practice, and a prepaid (capitated) multispecialty group practice. Ten conditions were chosen for review: periodic adult medical examinations, periodic gynecologic examination, periodic pediatric medical examination, therapeutic use of drugs with major side effects, anemia, hypertension, chronic heart disease, vulvovaginitis, acute urinary tract infection, and chronic urinary tract infections. Criteria for good performance were set by physicians representing the five sites. These studies showed that (a) specialists perform better but only in their specialized area of training, (b) younger physicians performed better, and (c) board certification was not consistently related to good performance.

Several other techniques were developed, primarily for assessing quality in ambulatory care.

The staging technique (Gonella et al., 1984) provides a method to characterize diseases according to how far they have progressed; the more advanced the stage, the less adequate prior care is assumed to have been. Judgments about adequacy of care are relative, depending on comparison of the stages of disease seen in one facility with those seen in another (Gonella et al., 1977).

The algorithm approach (Greenfield et al., 1981) is another technique suited to assessing the workup and management of problems as well as other aspects of care. In contrast with the other methods, the algorithm approach starts with the patient's problem rather than with the diagnosis. Flow diagrams start with a patient's problem (such as headache) and consist of a series of branchings depending on how subsequent questions from the patient's history and medical workup are answered.

In New York City, several hospital outpatient facilities have incorporated this approach into ongoing care. The use of these specified directions for diagnosis and management of children with respiratory problems, acute otitis media, asthma, and gastroenteritis results in better documentation in the medical record, fewer nonindicated laboratory procedures, more appropriate antibiotic prescribing, and reduced overall charges for services. Improved management, as measured by more rapid and more complete resolution of symptoms in the facility, has also been demonstrated, at least for asthma (Cook and Heidt, 1988).

In another approach, the *tracer* method (Kessner et al., 1973), patients are chosen from community surveys to detect individuals with certain types of problems. Their physician or physicians are identified and their process of care is traced to determine the recognition, diagnosis, and therapy of the problem as well as the follow-up. Although this method is too costly and impractical for routine use, it could be carried out on a sample basis when combined with other surveys such as the ongoing National Health Interview Survey. In this way, the costs of initially identifying patients with problems could be obviated, leaving only the costs of follow-up.

A similar approach was used on a large native Indian reservation in Arizona (Nutting et al., 1981, 1982). The process of care for each of nine conditions was traced from its prevalence in the community to screening procedures, through diagnostic evaluation, treatment and monitoring, to the institution of ongoing management where needed. Not all conditions were suitable for surveillance at each stage, for example, only ongoing management was useful for seizure disorders. For most of the conditions (hypertension, prenatal care, infant care including immunizations, lacerations, streptococcal pharyngitis, and gonorrhea) at least three phases of care could be assessed. Systematic differences in performance among different health service units on the reservation were noted. The advantage of tracer methods is their applicability to populations rather than only to patients in a facility; the assessment starts with the prevalence of a problem in the community and traces it through the processes of medical care to determine when and if poor quality of care exists.

Another approach to improving quality of care starts with the assumption that it is impossible for physicians to recall all the information required to provide optimum quality of care. To surmount this problem, a computer provides reminders to do certain things when certain situations exist, for example, a reminder to order a laboratory test if a patient is taking a drug that causes electrolyte imbalance (McDonald, 1976).

Outcome assessments are based on the assumption that high-quality care should result in improved health for the patient. In an early application of this method (Mushlin and Appel, 1980) patients were contacted several days after a visit to determine whether they still had symptoms, how much limitation of activity they were experiencing, and their understanding of the cause and prognosis of their problems. When these outcome assessments were compared with standard medical record audits, the assessments were more sensitive than the audits used to detect substandard care, at least as measured against overall expert judgments of the quality of care.

Judgments of the quality of care that are based on assessment of outcomes do not always agree with judgments made on the basis of a review of the characteristics of the care provided, even in the same patient. For example, Brook et al. used information from medical history questionnaires, screening examinations, insurance claims, and interviews with physicians and patients to judge the quality of ambulatory care (not necessarily limited to primary care physicians) for three chronic conditions in adults and four chronic conditions in children. Expert judges suggested that although approximately 70 percent of patients were receiving suboptimal care according to predetermined criteria, better care would result in more than minor improvement in fewer than half of the patients (Brook et al., 1990).

Since the purpose of medical care is to improve health and to relieve suffering, assessments that directly address these characteristics are of greater value than assessments of the procedures that are intended to achieve them. A large-scale study by 523 clinicians, which included family physicians, general internists, cardiologists, endocrinologists, psychiatrists, clinical psychologists, and other mental health professionals, examined the outcomes of care for adult patients with several chronic conditions. Outcomes included clinical measures such as blood pressure; physical, social, and role functioning of patients in everyday life; patients' perceptions of general health and well-being; and patients' satisfaction with treatment (Tarlov et al., 1989). One of the early findings from this study indicates that most of the differences in functioning and in well-being are not explained by the presence of specific chronic medical conditions in patients (Stewart et al., 1989) and that all health status measures were worse if patients also had depressive symptoms (Wells et al., 1989). These and other studies provide ample evidence that assessment of the outcomes of care for particular health problems must take into account the presence of other conditions ("comorbidity") experienced by patients.

There are many challenges to assessing quality of primary care. There is little doubt that access to care is important in maintaining a basic level of health. This impact has been demonstrated in children (Starfield, 1985) and in adults (Lurie et al., 1986; Fihn and Wicher, 1988). The remaining challenges are largely methodologic, and concern a number of issues such as the following:

- How can profiles of care be made more useful as a technique for improving quality? Are differences across facilities due to quality of care, or are they a result of differences in the severity of illness or greater burdens of illness (more comorbidity) among their patients?
- Can methods to assess the adequacy of physicians' recognition of patients' problems be developed and widely applied?
- How can assessments of outcomes be made more useful as a tool for improving the quality of care? Medical care cannot always prevent death, nor can it always avoid morbidity. To what extent can it be expected to minimize disability and enhance physical and psychosocial functioning? How can it maximize well-being in the present and potential for well-being in the future, and how can these aspects of health be measured with precision and reliability? Can a variety of types of outcomes be assessed so that methods to detect the impact of medical care will have greater sensitivity and specificity?

Improving Quality in Primary Care

In primary care, approaches to improving the quality of care may be generated from within clinical settings such as offices or group practices or it may be imposed by external groups acting as agents of the public.

Historically, efforts to ensure the quality of care have followed the state of the art of quality assessment. As the scientific basis for medical practice developed, the adequacy of physician training became an issue. Early in the twentieth century, reforms instigated by Flexner in the United States made education the focus of activities aimed at improving the quality of medical practice. For the first half of the century, the only systematic attempts to improve quality of care consisted of the credentialing of medical schools and instituting state requirements that graduates pass a standardized examination before being licensed.

Subsequent efforts to improve quality focused on other aspects of education. Studies have shown that the length of postgraduate training is an important determinant of quality of care (Palmer and Reilly, 1979), and for this reason some health care organizations require physicians to take examinations to qualify for certification by the "board," in their field of interest (board-eligible) or actually to have passed the examination for certification (board-certified). Participation in continuing education is required as a condition for renewal of licensure in only twenty-nine states and territories (twenty-seven states, the District of Columbia, and Puerto Rico), although the requirement is enforced in only twenty-three of these (Wentz et al., 1990). Some professional organizations in the United States, including the family physicians' organization, require a passing grade on a periodically administered test. Another approach useful for certain aspects of care, such as recommendation for surgery, requires a second opinion, that is, the patient must visit another physician who may concur or disagree with the initial recommendation.

The development of new techniques of evaluation and the availability of better information systems opened new possibilities for quality assurance. In the early 1970s legislation established the Professional Standards Review Organizations (PSROs). PSROs were organizations composed of physicians charged with assessing care in hospitals in their geographic area. At its maximum, the PSRO program had about two hundred local organizations spread over the country. Although they were limited to hospital care, the approach is applicable to outpatient care. PSROs conducted three types of review: *utilization review, medical care evaluations,* and *profiles of care.*

Utilization review was directed at regulating the amount of hospitalization, by reducing presumably unnecessary hospitalizations (preadmission certification), and by reducing the length of stay in hospital (continued stay review).

In *medical care evaluations* medical records are examined to determine if the patient's management was appropriate to the diagnosis. As the process of selection of records for review depends on the presence of a diagnosis in the record, the technique excludes patients who have the diagnosis but do not have it recognized or recorded by their physicians. Thus, medical care evaluations cannot address that component of the process of care that deals with the adequacy of problem recognition.

Profiles of care are statistical descriptions of care, as were the statistical descriptions used in the 1950s; where there were differences across hospitals, they triggered an investigation to determine whether the differences were medically justifiable.

The PSRO program was abandoned after several years for three major reasons: utilization review did not sufficiently reduce costs of hospital care to justify its own cost; medical care evaluations were never widely used, and sometimes their use increased the number of services and hence the cost of care; and profiles of care were rarely used, partly because it was not possible to attribute differences in performance solely to differences in the quality of care.

In the 1970s, attention shifted to the problem-finding method in which possible problems were noted and decisions were made to investigate them. Attention centered largely on administrative problems, such as scheduling appointments, rather than on clinical issues (Benson & Townes, 1990).

In the 1990s, attention has shifted to monitoring and evaluation, combining both medical care evaluations and problem evaluations, but with mechanisms that are better developed and more standard in their application.

Challenges to Quality Improvement in Primary Care

Dr S was lecturing to a class of freshman medical students on the subject of quality of care. Halfway through the lecture, a student in the back of the room raised his hand and said: "Dr. S, I don't know why we need a lecture on the quality of care. We had to be pretty smart to get into medical school in the first place and we all must be about the best there is since we were accepted to *this* medical school which is the best in the country. I don't think there is any question that the care we deliver will be of high quality, so why do we need a lecture on quality of care?"

Ensuring the quality of care is more of a challenge in the ambulatory setting than in the inpatient setting (Palmer, 1988). When the focus is on the episode of hospitalization, the outcome is more clearly defined; patients either die or are discharged and their status on discharge is relatively easily described. In the ambulatory setting, and particularly in the primary care setting, the "product" is difficult to define and even harder to measure. The functions of primary care are sometimes to cure, but are more often to care for patients until their problems resolve, stabilize, or at least stop progressing. In either case, or even when neither is possible, making patients comfortable and able to function to the limits of their capacity challenges the primary care professional.

In primary care many problems of patients are so poorly understood that the nature of their course or progression is unknown. Therefore it is difficult to determine what changes in conditions of patients are a result of therapy and which would have occurred in time without treatment.

In outpatient care, in contrast to inpatient care, contact with patients is limited to a few minutes, at infrequent intervals. Therefore, health personnel are less able to observe changes in illness and have less control over the management of treatment than they have in the inpatient setting.

In primary care, the unit of care is often undefinable. Many health problems have an uncertain onset, and primary care physicians encounter illnesses at different stages in different patients. In addition, the resolution of health problems is usually gradual, without a clear endpoint and some problems do not resolve at all but rather go through periods of waxing and waning. Time, rather than a visit, is a more useful unit of analysis for quality assurance in ambulatory than in inpatient care, although the appropriate period of time may be difficult to specify. In contrast, a hospitalization is a clear "episode," for which there exists a definable reason and which has a point of entry and a clear endpoint when the results of hospitalization can be described.

Primary care practices are usually much more isolated from peer influence than in hospital practice, so that problems of quality in primary care may be more chronic and ingrained than would be expected for hospital-based physicians. Furthermore, primary care data systems are not nearly as well developed as those in hospital care, despite major advances that have recently taken place. Ambulatory medical records are generally much less accessible and less complete than are hospital records.

These differences in primary care complicate the challenges of quality assurance, and approaches other than those used in hospitals are needed. Outcome measures of assessment that focus on levels of disability and mobility as well as on physical and mental functioning are at least as important outcomes of primary care as are disease resolution and improvement in biological manifestations of illness.

Measures that do not depend on specific diagnoses will complement those specific to particular diseases and may even be more useful for many purposes of quality assurance. These "generic" measures include characteristics such as (a) unusually low rates of recognition of certain types of problems commonly encountered in primary care practice, (b) apparently excessive hospitalization or referral rates which might indicate inadequate primary care, (c) excessive prescribing of dangerous or generally contraindicated medications, (d) delays in diagnosis of various types of problems that might suggest compromised access to services, or (e) adverse events such as unexpected deaths or emergency care for problems that proper care could have averted.

Sources of Information for Quality Improvement

Since primary care is increasingly covered by insurance plans or by governmental programs, the quality of care is a growing concern of these "third-party" payors. To maintain surveillance over quality of care, these agencies now require more information on forms previously used only for submitting reimbursement claims, information useful for some types of quality assurance activities. One problem with the use of claims forms is the uncertain accuracy and completeness of the data, which results in the need for periodic random checks of the information during the course of quality assurance activities. Computerized encounter forms serve the same purposes as claim forms and have similar limitations. However, they generally contain more information than claims forms, are easier to use, and are usually more available.

Several other techniques can facilitate quality improvement in the primary care setting. Periodic and systematic telephone surveys of patients are sources of information concerning adequacy of access to care, satisfaction with care, and responsiveness to medical therapies. Patients can be questioned about the time it took for abatement of symptoms and about difficulties they may have had with the prescribed treatment. Employment of simulated patients on a regular basis can provide information about the way care is delivered and whether patients had concerns about any aspects of it.

Judgments about the adequacy of services may be based on either standards set a priori or by comparisons with other facilities. When facilities are compared, differences in quality of care are often difficult to interpret because they might be due to differences in the quality of care or in the extent or severity of illnesses in the populations in the various facilities. This is a matter of concern whether the focus of attention is on specific diseases or is on generic measures of quality. In either case, comparisons of quality between two facilities are interpretable only if the extent of morbidity of their populations is the same. Some populations may be sicker than others and require a different type of care even for the same condition under investigation. If the degree of morbidity is not the same, the same criteria for adequacy of care may be inappropriate and cannot be applied to both. An approach of characterizing comorbidity in ambulatory settings is known as the ambulatory care group (ACG) method. This method characterizes patients according to the types of combinations of their health problems over a period of time. The ACG system predicts both concurrent and subsequent use of services and also distinguishes populations known to have different burdens of morbidity. It can also be used in studies of the variability in practice patterns across geographic area and among primary care physicians (Starfield, 1991). Techniques such as the ACG method are required in assessing the quality of care to characterize comorbidity so that comparisons among different facilities can be more accurately interpreted.

Areas of Concern for Quality Improvement

Although technically all aspects of primary care, including the attainment of its essential elements, can be subsumed under the rubric of *quality,* it is more consistent with common applications of the term to confine attention to the processes of care or to the resulting outcomes. Five considerations guide the selection of topics for quality assurance: (1) access to needed components of care, (2) technical quality of care, (3) appropriateness of care, (4) outcomes of care, and (5) satisfaction with care.

Access. Access to hospital care is sometimes problematic, and may be particularly difficult if there are geographic or financial barriers. It is generally assessed by surveying patients, by simulated patients, or by examining characteristics of the facility that might impede entry. Certain aspects of access can also be assessed by reviewing medical records and claims forms. In the context of quality of the process of care, *access* refers to the performance of needed and indicated procedures. For example, immunizations that are delayed or not received at all, failure to receive mammograms or Pap tests, failure of adults to know their blood pressure, or excessive delay in seeking care in the presence of symptoms are indicators of poor access to care, since receipt of an adequate amount of services should ensure that needed care is obtained in a timely manner.

Technical Quality. Technical quality consists of the attainment of standards of care for recognition of patients' needs, diagnostic procedures and decisions, therapy and management of diseases and disorders, and strategies for reassessment. For example, patients under care for diabetes should have adequate monitoring of parameters reflecting control of their blood sugar, children with asthma should not routinely be receiving oral cortiosteroids for long-term management of their asthma, and treatment for otitis media should include a follow-up to ensure that the inflammatory process has healed and hearing has returned to normal. In most cases, inappropriate care results from errors of omission when indicated procedures were not undertaken. Sometimes, however, errors of commission from faulty judgment result in wrong procedures. Technical quality is generally assessed by reviewing medical records, abstracts of medical records, encounter forms, or claims forms.

Appropriateness. Appropriateness concerns the performance of procedures that, according to current standards, are *not* justified and result in too much rather than too little being done for the patient. An important focus of current quality assurance activities involves hospitalizations and procedures that appear to be unwarranted. Information concerning appropriateness is generally obtained by comparing rates of procedures or hospitalizations across facilities or populations. Higher rates suggest a suspicion of unnecessary procedures. Appropriateness is also ascertained by reviewing proposed care to determine its justifiability. For example, meeting requirements for preadmission certification of proposed hospitalizations can ensure that unnecessary ones are avoided. Preadmission certification is usually used as a cost-control mechanism, but the principle pertains to quality assurance as well. Examples of inappropriate care include the performance of caesarian sections or tonsillectomies when criteria for these procedures are not met.

Outcomes. Outcome assessments comprise approaches that assess the health status of individuals after medical care. Avoidance of premature death, improved health, and adequate preventive and health promoting activities are examples of types of measures of quality assurance that concern the outcomes of care. Information on outcomes is not uniformly present in medical records; what information is present may not reflect the adequacy of care in all patients because not all may return for visits after therapy is prescribed. Therefore, information about outcomes of care is usually obtained from patients through telephone calls, mail questionnaires, or formal surveys. Where populations are large, certain types of information might be obtainable from community data such as birth certificates (for birth weight), death certificates (for rates and cause of death), preschool surveys (for immunization rates), and hospital discharge summaries (for rates of hospitalization by cause and lengths of stay).

Patient Satisfaction. Certain aspects of the quality of care are ascertained by periodic assessment of patients' satisfaction with services. Necessary information comes from surveys conducted by telephone, mail, or in person. To avoid bias engendered by questioning only those patients who come to a facility, it is necessary to select individuals randomly from a patient register, to reach those who might be so dissatisfied they avoid contact with the health service.

Non-clinical Characteristics. Characteristics other than those relating to physician–patient interactions are also amenable to quality improvement. These include the capabilities of the professional and support staff, mechanisms to ensure continuity of care and adequacy of medical records, and characteristics of the facility that influence patient safety and comfort. Standards for continuity of care can be set and monitored by reviewing appointment log books or medical records. Audits of the latter ensure that the required minimum of information is present in forms such as problem lists, medication lists, laboratory reports, and maintenance of recommended format for patients' histories and other data. Regular surveys of the facility will ensure the presence of appropriate safeguards. The credentials of personnel should be investigated at employment and updated periodically to ensure that continuing education requirements are met. The three major determinants of quality of care are the length of the physician's experience in managing the particular problem for which quality is assessed, the nature of the organization in which physicians work, and the length of postgraduate training (Palmer and Reilly, 1979). These determinants merit consideration in every quality improvement program.

Clinical Approaches to Quality Improvement

The Joint Commission on the Accreditation of Healthcare Organizations (JCAHO) recommends that certain guidelines be followed in designing an ambulatory quality assurance activity for facilities that wish its accreditation (JCAHO, 1990).

Quality assurance activities involve two related phases: monitoring and evalua-tion and problem resolution. Nine steps are the basis of these phases (Benson and Townes, 1990):

- Determining the focus of responsibility for the activities.
- Determining the scope of services in the activity.
- Selecting specific aspects of care to be tracked routinely.
- Choosing specific indicators of quality to monitor.
- Developing criteria for evaluating performance and standards within these criteria that should be met.
- Routinely and systematically monitoring and evaluating the chosen indicators by col-lecting and analyzing data and comparing them against the preset standards.
- Taking appropriate actions to resolve identified problems.
- Tracking corrective actions to ensure that they are operating as planned.
- Creating a permanent record of all quality assurance activities to ensure that all infor-mation is integrated into the organization.

Since claims forms are easily available they are a logical source of information about several aspects of the quality of care. Data from claims forms, especially when linked to "beneficiary files" to obtain information on demographic charac-teristics, are useful for quality improvement activities in enrolled populations such as Medicare recipients or patients in a health care organization. Weiner et al. (1990) recommended sixty indicators of quality that employ information obtained from claims forms and that one can use to monitor quality of care in such populations.

One category of these indicators reflects preventive care; it includes procedures that prevent disease and screening procedures that detect illness at an early stage. An example of this category is the percentage of children or adults receiving recom-mended immunizations at the appropriate time. Examples of criteria for screening include the percentage of persons receiving a recommended test (such as Pap test for cervical cancer for adult women) and the percentage of infants receiving the recom-mended number of well-child visits.

A second category of indicators concerns the adequacy of diagnostic pro-cedures. One example is the percentage of persons with a given diagnosis who received the recommended diagnostic tests; another is the overuse of procedures or medications that needlessly expose individuals to iatrogenic problems.

A third category concerning treatment and management has the largest number of suggested indicators. Among the indicators are those concerning medications such as rates of patients receiving inappropriate or contraindicated medications, and those concerning therapeutic surgical procedures such as the rates of tonsillectomy or the extent to which criteria for performing tonsillectomies are met. Also included within this category are follow-up procedures, such as the percentage of patients lost to follow-up. Indicators of continuity or of access to care include such elements as the percentage of all visits without a referral made to the patient's primary care provider, the percentage of patients who experienced delays in receiving indicated procedures, and the percentage of visits for emergency care that earlier contact with the primary care provider could have averted. Specialty referrals is another type of indicator in this category, one example is the percentage of patients over the age of fifty with insulin-dependent diabetes who should be regularly examined by an ophthalmologist.

Yet another category of indicators are those which reflect general systems functions such as rates of hospitalization, readmissions, and sentinel conditions, as discussed in Chapter 13. Additional categories of indicators reflect inpatient care that may be under the control of the primary care physician.

Some of these indicators reflect the processes of care and some reflect outcomes. Some are related to access to care, others to technical quality, and yet others to appropriateness of care. High or low rates of occurrence of indicators are the basis for interpreting some of the criteria while for others the occurrence of specific untoward events signal a problem with quality. Indicators such as preventive activities reflect desirable events while others represent undesirable ones.

When information about quality of care is obtained from claims forms, it is limited to users of services who are not necessarily representative of those eligible to receive care. Another limitation of claims records is the relative paucity of information concerning details of clinical care. Like medical records, these are of little use in assessing the extent to which practitioners adequately recognize the concerns that bring patients to seek care. This latter facet of care is not yet a feature of most quality improvement activities. Assessment of other aspects of diagnosis, therapy, and reassessment requires additional information from sources other than claims forms to identify the nature and source of the lack of quality so that remedial steps can be undertaken. Surveys of patients, in which samples are selected from enrollee rosters rather from those who visit facilities, are more suitable to population-based assessments that involve people eligible for care who do not use it as well as those who do.

Most methods that address quality of care are based on the assumption that there are both right and wrong ways to practice and that information can distinguish between them provided the correct information is available. Another method, the *theory of continuous improvement,* is based on other assumptions (Berwick, 1989). This theory assumes that the level of knowledge is imperfect, that no one performs perfectly, and that everyone is motivated to improve, given the opportunity to do so. The assumption holds that setting "minimal" standards of care thwarts the process of long-term quality improvement which is based on new knowledge about the delivery of care and the participation of those who deliver it. By this method health professionals who deliver care help generate the topics of concern and are involved in seeking the reasons for and possible solutions to inadequate performance. Although this approach does not eliminate the need to develop and apply tests of truly inadequate performance, it does shift the emphasis to a process that involves everyone in more constructive thinking and planning activities to improve the quality of care.

The Quality of Care and Technologies (QCT) program of the World Health Organization in Europe is based on these principles. The program's main functions are as follows:

- To establish mechanisms for the assessment and assurance of the quality of health care through studies and projects aimed at the production of guidelines, criteria, and standards.

- To ensure the development and appropriate use of information systems to be used for the assessment and improvement of health care.
- To identify tracer conditions and models in hospital and primary health care quality assurance programs.
- To coordinate national and intercountry research studies of health services.
- To collaborate with nongovernmental and governmental organizations and national institutions in the field of health technology assessment and utilization, and to use the results to develop quality of care.
- To identify institutions able to collaborate in promoting, implementing, and developing quality assurance in health care services.
- To organize national and international courses and seminars for promoting and developing quality assurance concepts and methods.

The program actively promotes the appropriate use of computerized information systems as a tool to improve quality and outcomes in primary health care, organizes workshops and consultations to explore reasons for variations in health care practice, and designs and coordinates research projects to investigate reasons for these variations and to assess the effectiveness of various technologies and medication use in primary care. Examples of its activities include studies on the appropriate use of iron during pregnancy, the effectiveness of diagnostic and treatment strategies for acute tonsillitis, and various aspects of the management of diabetes mellitus (World Health Organization, 1990).

Regulatory Approaches to Quality Assurance

Growing concern about the usefulness of many medical interventions and the increasing availability of techniques to document and compare them in different settings and geographic areas has stimulated the development of activities on a scale larger than that in individual clinical settings.

To initiate quality assurance activities on a wide scale, there must be guidelines to which practitioners and health services organizations can be held accountable. The setting of guidelines is not, by itself, a new activity, since for many years professional groups have engaged in setting standards for a variety of medical interventions. For example, the American Academy of Pediatrics publishes a series of manuals which provide guidance to practitioners as to the components of well-infant and well-child care and the appropriate diagnosis and management of infectious diseases in children (American Academy of Pediatrics, 1988a, 1988b). In addition, large health insurers, such as the Blue Cross and Blue Shield Associations, have been attempting to set guidelines for the appropriateness of many diagnostic and treatment services.

In 1977 consensus conferences convened by the National Institutes of Health involved a selected panel to review existing evidence and recommend procedures to be followed for specific interventions, such as caesarian sections (Perry, 1987). Similar panels of experts convened by a variety of professional groups are judging the appropriateness of a whole variety of medical and surgical procedures (Park et al., 1986), although current efforts are largely limited to inpatient rather than to outpatient or primary care practices. One set of guidelines is particularly relevant to primary care physicians since it presents the criteria and justification for a variety of screening procedures commonly used in primary care (U.S. Preventive Services Task Force, 1989). Data from such efforts are being used in negotiations for payment under National Health Service contracts in Great Britain as well as for payments by government-funded programs in the United States (Bunker, 1990).

As the number of sets of guidelines and standards proliferates, it becomes apparent that not all are in agreement, even when the subject under consideration is the same. Moreover, many important areas remained neglected, and there are great gaps in public disclosure, not only of the guidelines themselves but also of the process used to formulate them. There are conflicts in terminology and in technique as well as a relative lack of attention to systematic implementation of recommendations (Field and Lohr, 1990).

In recognition of these deficiencies, the U.S. Congress created an Office of the Forum for Quality and Effectiveness in Health Care. This forum in the Agency for Health Care Policy and Research (AHCPR), a component of the Department of Health and Human Services, must arrange for the development, periodic review, and updating of mechanisms for quality assurance. Two related facets are specified under the legislation.

- Clinically relevant guidelines that may be used by physicians, educators, and health care practitioners to assist them in determining how diseases, disorders, and other health conditions can most effectively and appropriately be prevented, diagnosed, treated, and managed clinically.
- Standards of quality, performance measures, and medical review criteria through which health care providers and other appropriate entities may assess or review the provision of health care and ensure the quality of such care (IOM, 1990).

The AHCPR is charged not only with developing guidelines but also with supporting and encouraging research efforts to provide the scientific basis for development of guidelines. The Medical Treatment Effectiveness Program (MEDTEP) starts with a format for effective research that includes a literature review and synthesis, an analysis of variations in medical practice and patient outcomes, and dissemination and an evaluation of effectiveness findings (AHCPR, 1990). Although the initial efforts are directed at care of the elderly because of the federally financed Medicare program, it is likely that care of younger individuals and care in the primary care sector will follow.

Government-mandated quality assurance responsibilities are vested in Peer Review Organizations (PROs) created by the U.S. Congress in 1982. The Health Care Financing Administration (HCFA), an agency of the Department of Health and Human Services, grants contracts based largely in individual states to ensure that care is medically necessary, rendered in the most appropriate and cost-effective manner, and meets professionally recognized quality standards. PROs are responsible for monitoring the documentation in the medical record to determine if hospitals and physicians are achieving these goals. Initial PRO activities focused on the inpatient setting and only on patients covered by the Medicare program; payment was denied if unnecessary hospitalizations, hospital days, or procedures were found. In 1985 the provision of the Consolidated Omnibus Reconciliation Act added the following responsibilities to the PROs:

- Review of ambulatory surgery procedures performed in hospital outpatient departments and in ambulatory surgical centers.
- Review of intervening care provided by home health agencies, skilled nursing facilities, and hospital outpatient departments.
- Preadmission/preprocedure review of ten preselected procedures to determine their necessity and the appropriateness of the setting.

PRO activities are initiated by nurses trained to screen medical records for certain characteristics based on physician-developed criteria. If care does not meet the criteria, the case is referred for peer review by either an individual physician adviser or a panel of physician specialists. If care is found to be unwarranted, the PRO notifies the physician and hospital of the identified problem and requests additional information. A PRO physician reviews this and decides if disapproval is indicated. If so, payment is denied and any payment already disbursed must be returned. In reviews of quality, the PRO recommends corrective action to ensure that the problem does not recur.

In the PRO setup, problems of medical quality are of three types: mismanagement without the potential for significant adverse effects on the patient; mismanagement with the potential for significant effects on the patient; or mismanagement with significant adverse effects on the patient.

All confirmed quality problems are assigned points and quality profiles of hospitals and physicians are established. Indicated remedial interventions include notification by letter, telephone, or in-person discussions, literature references, continuing medical education courses, intensified review, notification of licensing and accreditation agencies, or sanctions. Certain types of PRO information must be kept confidential, including the following (Delmarva, 1990):

- Information that explicitly or implicitly identifies an individual patient, practitioner, or reviewer.
- Sanction reports and recommendations.
- Quality review studies that identify patients, practitioners, or institutions.
- PRO deliberations.

Within the near future, PROs are likely to become involved in reviewing ambulatory as well as inpatient care. Some states already require PROs to review care rendered to patients in state Medicaid programs for the medically indigent.

As a result of the Health Care Quality Improvement Act of 1986, professional review bodies acting under the authority of existing federal legislation now have legal immunity from prosecution. The law also created reporting requirements under which any formal actions against physicians or hospitals, including lawsuits or insurance settlements, must be reported and maintained in a national data bank. Accrediting bodies and hospitals are obliged to consult the bank before they grant professional privileges to physicians (Igelhart, 1987).

Special difficulties in specifying and measuring quality of care in the primary care setting are likely to interfere with the early application of large-scale formal quality improvement mechanisms. However, certain types of primary care organizations, such as HMOs, could contribute to the development of mechanisms to facilitate their use (Weiner, 1986). One area concerns the application of information systems that lend themselves to quality improvement activities. HMOs are more likely to have information systems that are relatively accessible and relatively complete, at least compared with records in most offices of individuals or small groups of physicians. HMOs and other practices that now involve patients in decision-making or in an advisory capacity could also develop prototypes for consumer participation in quality assurance. Primary care organizations with defined populations might also experiment with techniques to combine population-oriented measures with facilities-based measures to provide an improved basis for planning and policy formulation.

Public disclosure of information is an approach to quality care that will be increasingly used. The assumption is that providing consumers with information about the performance of health care organizations will help them select physicians or facilities whose performance is better, thus encouraging all physicians and facilities to provide superior service. This potential approach was recognized by the Institute of Medicine (IOM, 1981), which stated that "the feasibility and effectiveness of publicizing instances of persistent poor quality by individual practitioners should be explored." In its subsequent report on confidentiality of hospitalization data, the IOM endorsed the policy of disclosing to the public all information except that which identifies specific patients. It recommended that information be available in a form in which institutions are identified, and in a form that identifies specific physicians by unbreakable code but not by name. It also recommended that quality review organizations upon request disclose to a patient or a patient designate any data about the patient derived from a medical record abstract (Institute of Medicine, 1981). Current applications of this method focus on mortality rates in hospitals. Further development of approaches to specifying appropriate outcomes of ambulatory care could lead to public disclosure concerning this aspect of the performance of primary care physicians as well.

Thus, the most recent approaches to quality improvement increasingly involve patients not only in judging their own care but as active participants in choosing their place of care on the basis of information about its prior performance.

References

American Academy of Pediatrics. Guidelines for Health Supervision, 2nd ed. Elk Grove Village, Ill., 1988.

American Academy of Pediatrics. Report of the Committee on Infectious Diseases, 22nd ed. Elk Grove Village, Ill., 1991.

AHCPR (Agency for Health Care Policy and Research). Medical Treatment Effectiveness Research. Program Note, U.S. Department of Health and Human Services, Rockville, Md., March 1990.

Benson D and Townes P. Excellence in Ambulatory Care. A Practical Guide to Developing Effective Quality Assurance Programs. San Francisco, Jossey-Bass Publishers, 1990.

Berwick D. Continuous improvement as an ideal in health care. N Engl J Med 1989; 320:53–56.

Brook R, Kamberg C, Lohr K, Goldberg G, Keeler E, and Newhouse J. Quality of ambulatory care: epidemiology and comparisons by insurance status and income. Medical Care 1990; 28:392–433.

Bunker J. Variations in hospital admissions and the appropriateness of care: American preoccupations? Br Med J 1990; 310:531–32.

Cook CD and Heidt J. Assuring Quality Out-Patient Care for Children. Guidelines and a Management System. New York, Oxford University Press, 1988.

Delmarva Foundation for Medical Care Inc. Know Your PRO, 1990.

Eisele C, Slee V, Hoffmann R. Can the practice of internal medicine be evaluated? Ann Intern Med 1956; 44:144–61.

Fihn S and Wicher S. Withdrawing routine outpatient medical services: effects on access and health. J Gen Intern Med 1988; 3:356–62.

M Field and K Lohr (eds). Institute of Medicine (IOM). Clinical Practice Guidelines. Directions for a New Program. Washington, D.C., National Academy Press, 1990.

Gonnella J, Cattani J, Louis D, McCord J, and Spirlca C. Use of outcome measures in ambulatory care evaluation. In: Giebink G, White N, Short E. Ambulatory Medical Care-Quality Assurance 1977. Proceedings of the Conference, La Jolla Ca, La Jolla Science Publications, 1977.

Gonnella J, Hornbrook M, and Louis D. Staging of disease: a case-mix measurement. JAMA 1984; 251:637–44.

Greenfield S, Cretin S, Worthman L, Dorey F, Solomon N, and Goldberg G. Comparison of a criteria map to a criteria list in quality-of-care assessment for patients with chest pain: the relation of each to outcome. Medical Care 1981; 19:255–72.

Horn SD, Buckle JM, and Carver CM. Ambulatory severity index: development of an ambulatory case mix system. J Amb Care Manage 1988; November:53–62.

Igelhart J. Health policy report. Congress moves to bolster peer review: The Health Care Quality Improvement Act of 1986. N Engl J Med 1987; 316:960–64.

Institute of Medicine. Access to Medical Review Data. Washington, D.C., National Academy Press, 1981.

JCAHO (Joint Commission on Accreditation of Healthcare Organizations). Ambulatory Health Care Standards Manual. Chicago, 1990.

Kessner D, Kalk C, and Singer J. Assessing health quality—the case for tracers. N Engl J Med 1973; 288:189–94.

Lembcke P. Medical auditing by scientific methods. JAMA 1956; 162:646–55.

Lurie N, Ward N, Shapiro M, Gallego C, Vahaiwalla R, and Brook R. Termination of Medi-Cal benefits. A follow-up study one year later. N Engl J Med 1986; 314:1266–68.

McDonald C. Protocol-based computer reminders, the quality of care and the non-perfectibility of man. N Engl J Med 1976; 295:1351–55.

Morehead M and Donaldson R. Quality of clinical management of disease in comprehensive neighborhood health centers. Medical Care 1974; 12:301–15.

Mushlin A and Appel F. Testing an outcome-based quality assurance strategy in primary care. Medical Care 1980; 18(5;Suppl):1–100.

Nutting P, Shorr G, and Burkhalter B. Assessing the performance of medical care systems: a method and its application. Medical Care 1981; 19:281–96.

Nutting P, Burkhalter B, Dietrick D, and Helmick E. Relationship of size and payment mechanism to system performance in 11 medical care systems. Medical Care 1982; 20:676–90.

Palmer RH. The challenges and prospects for quality assessment and assurance in ambulatory care. Inquiry 1988; 25:119–31.

Palmer RH and Reilly M. Individual and institutional variables which may serve as indicators of quality of medical care. Medical Care 1979; 18:693–717.

Park R, Fink A, Brook R, Chassin M, Kahn K, Merrick N, Kosecoff J, and Solomon D. Physician ratings of appropriate indications for six medical and surgical procedures. Am J Public Health 1986; 76:766–72.

Payne B, Lyons T, and Neuhaus E. Relationships of physician characteristics to performance quality and improvement. Health Services Res 1984; 19:307–32.

Perry S. The NIH consensus development program: a decade later. N Engl J Med 1987; 317:485–88.

Starfield, B. Effectiveness of Medical Care: Validating Clinical Wisdom. Baltimore, The Johns Hopkins University Press, 1985.

Starfield B, Weiner J, Mumford L, and Steinwachs D. Ambulatory care groups: a categorization of diagnoses for research and management. Health Services Res 1991; 26(1):53–74.

Stein R, Perrin E, Pless IB, Gortmaker S, Perrin J, Walker DK, and Weitzman M. Severity of illness: concepts and measurement. Lancet 1987; (Dec 26):1506–09.

Stewart A, Greenfield S, Hays R, Wells K, Rogers W, Berry S, McGlynn E, and Ware J. Functional status and well-being of patients with chronic conditions: results from the medical outcomes study. JAMA 1989; 262:907–13.

Tarlov A, Ware J, Greenfield S, Nelson E, Perrin E, and Zubkoff M. The medical outcomes study: an application of methods for monitoring the results of medical care. JAMA 1989; 262:925–30.

U.S. Preventive Services Task Force. Guide to Clinical Preventive Services. Baltimore, Williams & Wilkins, 1989.

Weiner J, Powe N, Steinwachs D, and Dent G. Applying insurance claims data to assess quality of care: a compendium of potential indicators. Quality Review Bulletin, 1990; 16:423–38.

Weiner, J. Assuring quality of care in HMOs: past lessons, present challenges and future directions. GHAA Journal 1986; Spring:10–27.

Wells K, Stewart A, Hays R, Burnam A, Rogers W, Daniels M, Berry S, Greenfield S, and Ware J. The functioning and well-being of depressed patients: results from the medical outcomes study. JAMA 1989; 262:914–19.

Wentz D, Gannon M, and Osteen A. Continuing medical education. JAMA 1990; 264:836–40.

World Health Organization, Regional Office for Europe. Facing the Future: Quality Assurance. Copenhagen, Denmark, 1990.

13

Community-Oriented Primary Care

In this chapter, we come full circle back to the concepts of primary care expressed in the Declaration of Alma Ata (see Chapter 2). In the paragraph defining primary care, the word "community" was used three times: "at a cost the community can afford to maintain," "an integral part of . . . . the overall social and economic development of the community," and "the first level of contact of individuals, the family and community with the national health system. . . ." The declaration recognized the distinction between primary care as delivered to individuals or families and as delivered to the community; it required attention to both.

The presentation of the four components of primary care in chapters 3–6 explicitly distinguished methods that are useful on a population level from those that are useful mainly on a clinical level. However, most of the discussion and evidence of their usefulness was drawn from experiences in clinical settings. The reason for this is straightforward: little has been done to explore the usefulness or benefits of community-oriented primary care.

Traditional primary care—that is, care from the perspective of the clinician, who has been exposed exclusively to patients appearing for care—has evident limitations. Such care cannot take into consideration the distribution of health problems in the community, since many of those problems may never come to the clinician's attention, and even if they do the relative frequencies of problems may not reflect their relative frequencies in the community. In addition, traditional primary care cannot be aware of the way problems initially are manifested because patients often wait for problems to progress or change before they seek care. Further, traditional clinically oriented primary care has difficulty in understanding the relative impact of environmental, social, and behavioral factors in disease etiology and progression since practitioners are often unfamiliar with the milieu in which patients live and work. Finally, monitoring the impact of health services is also difficult in traditional primary care because follow-up or feedback requires that the patient contact or revisit the practitioner.

There are several reasons why a population-based focus in clinical care is desirable. The first is that knowledge about the distribution of health problems cannot be derived from experiences in medical centers or in individual practitioners' practices. Work done by White (1961) conclusively demonstrated that most people who experience an illness (about three-quarters of the population in a given year) do not consult a physician; a maximum of a third seek help. Therefore, information about the nature and distribution of health problems cannot and should not be based on the experiences of practitioners in medical centers or offices. The second reason is that knowledge of how disease presents itself is not obtainable without a population-based focus. Evidence on this point derives from a study in which physicians were asked to indicate the usual presenting findings for several genetic metabolic diseases (Holtzman, 1978). The physician generally cited the classic signs of disease as described in standard textbooks, which in fact are seldom the problems for which attention is sought.

The third reason for a population-based focus is that physicians overestimate their roles in providing care. Physicians report that most of their patients are "regular," yet a substantial minority of patient-initiated visits are made elsewhere (Dutton, 1981). When practitioners do not recognize this, as is usually the case (Starfield et al., 1976), both the attainment of the features of primary care and its quality are jeopardized.

The fourth reason concerns the importance of feedback for the continuing education of the practitioner. When patients fail to return to the practitioner for follow-up, either because they were not satisfied with the care or they were not happy with the results, the practitioner loses information critical in learning from experience. A community focus would enable practitioners to develop information systems that could monitor the loss of patients to their care and to assess reasons for it.

The final reason for a population-based focus for clinical care derives from the imperative to add new knowledge about diseases, their natural course in the absence of treatment, and how that course can be modified by various interventions. Most of what is known about illness and especially its management results from experiences with patients in medical research centers. Although community interview surveys obtain information concerning peoples' perceptions of their problems, and population examination surveys provide additional information about aspects of health that can be detected by examinations and laboratory tests, neither approach can link medical care interventions with health status, because clinical data are generally not available or linkable to survey data. Researchers in medical centers usually focus their attention on particular diseases. However, professional knowledge about disease does not necessarily reflect people's illness experiences. As medical care becomes more and more effective in reducing mortality from specific causes, people will survive and thus be at risk of multiple diseases which may interact in unknown ways and respond differently to various interventions. The disease-focused approach that characterizes most of clinical research will increasingly have to be supplemented by a person-focused one, and such a focus can be achieved only by viewing the individual in a community context as well as in the clinical facility.

What Is Community-Oriented Primary Care?

Community-oriented primary care (COPC) has been defined in many ways. Common to all is the notion that it is an approach to primary care that uses epidemiologic and clinical skills in a complementary fashion to tailor programs to meet the particular health needs of a defined population. It explicitly recognizes the interactions in the diagram of the determinants of health, as presented in Chapter 1; these include the overlap of the health services system and the social and physical environment as well as the overlap of the health services system and individual behaviors that influence health.

A community-oriented approach applies the methods of clinical medicine, epidemiology, social sciences, and health services research and evaluation to the following tasks (Nutting and Connor, 1986):

- Defining and characterizing the community.
- Identifying community health problems.
- Modifying programs to address these problems.
- Monitoring the effectiveness of the program modifications.

Applying epidemiologic methods where data are more representative than those derived from clinical practices should especially improve certain aspects of care. Diagnosis and management would be more appropriate because of better recognition of etiologic factors, many of which arise from social and environmental exposures. "Problem recognition" should also improve, as more complete data will provide information on the characteristics of early stages of illness; more complete data will also facilitate the recognition of new types of disorders and clusters of unusual symptomatology. Definitions of "normal" health can also be refined with more complete descriptions of the characteristics of individuals who rarely appear for care.

The application of social sciences techniques should improve the recognition of existing problems through understanding of the impact of social and economic factors on health, including those of poverty, unemployment, and other stressful states.

Application of health services research techniques would provide a better understanding of the impact of various aspects of medical care, and of the relationships between components of the structure, process, and outcomes of health services.

In approaches to measuring the attainment of primary care, community orientation is considered a "derivative" feature, in the sense that it would derive from a high level of attainment of the unique features of primary care. In the process of optimally achieving a high degree of longitudinality of care, a program would have defined the population eligible for care and, in addition, would have made sure that the population knew the program assumed responsibility for its health services. The program would thus achieve the first functional step of COPC, that of defining and characterizing the community in such a way that nonusers of services are not systematically excluded.

The second step in achieving COPC comes when the primary care program, in its quest for ultimate comprehensiveness, identifies community problems and recognizes the health needs of its enrolles, including those who rarely appear for care. The third and fourth steps—modifying the health care program and monitoring the effectiveness of the modification—would follow from the two derivative steps involving longitudinality and comprehensiveness.

TABLE 13.1.   Staging Criteria for COPC Function—
Defining and Characterizing the Community

Stage 0	No effort has been made to define or characterize a community beyond the active users of the practice.
Stage I	There is no enumeration of the individuals who comprise the community. The community is characterized by extrapolation from large area census data.
Stage II	There is no enumeration of the community, but it is characterized through the use of secondary data that correspond closely to the community for which the practice has accepted responsibility.
Stage III	The community can be enumerated and is actively characterized through the use of a data base that includes all members of the community, and that contains information to describe its demography and socioeconomic status. (Often such a data system is constructed over time from the active users of services, but approximates the community closely, e.g., at or above 90 percent coverage of the community.)
Stage IV	Systematic efforts ensure a current and complete enumeration of all individuals in the community, including pertinent demographic and socioeconomic data. For each individual, information exists that facilitates targeted outreach, e.g., address, telephone number, etc.

Source: IOM, April 1984.

The conceptual bases for COPC dates back to the writings of Will Pickles in Great Britain (Pickles, 1938); Sidney Kark further developed the concept, first in South Africa and then later in Israel (Kark, 1974, 1981). Several models have been evolving in the United States, although not as a formal movement toward COPC. In 1982, the Institute of Medicine of the National Academy of Sciences sponsored a conference and a study of COPC as it existed at that time in the United States.

To describe the degree to which COPC has been attained, the Academy specified the stages of achievement of each of the four aspects of COPC. Tables 13.1 through 13.4 describe the staging. Table 13.1 contains the stages in defining the community. At the lowest level, no effort has been made. At the highest level, there are systematic efforts that ensure a current and complete enumeration of all individuals in the community, including pertinent demographic and socioeconomic data.

Table 13.2 list the stages in identifying community health problems. At the lowest level no systematic efforts are made to understand the health status or health needs of the community; the results from studies of the patient population are assumed to reflect the health problems in the community as a whole. At the highest level there are formal mechanisms to identify and set priorities among a broad range of potential health problems in the community, to determine their correlates and determinants, and to characterize the existing patterns of health care related to the problems.

Table 13.3 presents the stages in modifying the health care program and extends from no modifications made in response to the needs of the community (Stage 0) to Stage 4, in which modifications in the program involve both primary care and community or public health components and are targeted to specific high-risk groups with active efforts to reach these groups.

TABLE 13.2.   Staging Criteria for COPC Functions—
Identifying Community Health Problems

Stage 0	No systematic efforts have been made to understand the health status or health needs of the community; the results from studies of the patient population are assumed to reflect the health problems in the community as a whole.
Stage I	Community health problems are identified through general consensus of the providers and/or community groups.
Stage II	Community health problems are identified by extrapolation from systematic review of secondary data, such as vital statistics, census data, large area epidemiological data, etc.
Stage III	Community health problems are examined through the use of data sets specific to the community, but perhaps focusing on single health problems or health care issues.
Stage IV	Formal mechanisms (usually but not always epidemiologic techniques) are used to identify the set priorities among a broad range of potential health problems in the community, identify their correlates and determinants, and characterize the existing patterns of health care related to the problem.

Source: IOM, April 1984.

Table 13.4, which lists the stages in monitoring the effectiveness of modifications in the primary care program, has five levels. At the lowest level, examination of program effectiveness is limited to the impact on the active users of health services. In the highest stage, the effectiveness of the program is determined by techniques that are specific to program objectives, account for differential impact among risk groups, and provide information on positive and negative impacts of the programs. Intermediate stages of achievement of these four functions describe progress toward attaining the highest level of the function.

TABLE 13.3.   Staging Criteria for COPC Functions—
Modifying the Health Care Program

Stage 0	No modifications are made in the primary care program in specific response to health needs of the large community.
Stage I	Modifications address health problems believed to exist in the community, but are made more in response to a national or organizationwide initiative than to a particular problem specifically identified within the commmunity.
Stage II	Modifications address important community health problems, but are chosen largely for the availability of special resources to address the particular problem, and closely follow guidelines that may not be tailored to the community needs.
Stage III	Modifications in the health care program are tailored to the unique needs of the community and involve (where appropriate) both the primary care and the community/public health components of the program.
Stage IV	Modifications in the program involve both primary care and community/public health components and are targeted to specific high risk or priority groups, with active efforts (e.g., outreach) made to reach specific high-risk or priority groups within the community.

Source: IOM, April 1984.

TABLE 13.4. Staging Criteria for COPC Functions—
Monitoring the Effectiveness of Program Modifications

Stage 0	Examination of program effectiveness is limited to the impact on the active users of health services.
Stage I	Program effectiveness is viewed in terms of impact on the community as a whole, but is based on subjective impressions of the practitioners and/or community groups.
Stage II	Program effectiveness is estimated by extrapolation from large area data or vital statistics.
Stage III	Program effectiveness is determined by systematic examination of a data set that is specific to the community.
Stage IV	Program effectiveness is determined by techniques that are specific to the program objectives, account for differential impact among risk groups, and provide information on the positive and negative impacts of the program.

Source: IOM, April 1984.

Achieving Community-Oriented Primary Care

The IOM study identified and characterized seven facilities seeking to provide COPC. Although none of the programs had achieved a high level of COPC, several had attained relatively high levels of performance on some of the four functions (Nutting and Connor, 1984).

A notable attempt to achieve COPC in clinical practice has been made by Dr. John Tudor Hart in Great Britain. Dr. Hart, whose practice is located in a Welsh mining town, takes responsibility for both community and clinical functions. In his concept, the "community general practitioner is a new type of physician who is engaged in local participatory democracy in the pursuit of the maximization of health" (Hart, 1983).

Mant and Anderson (1985) proposed that general practitioners accept responsibility for auditing the state of health of their patients and for publicizing the results, monitoring and controlling environmentally determined disease, auditing the effectiveness of preventive programs, and evaluating the effects of medical intervention. In Great Britain, responsibility for community functions is vested in a cadre of community medicine specialists. Mant and Anderson propose that the functions be assumed by the primary care practitioners, with transfer of resources from the current community medicine structure to the primary care practices.

In general, medical schools throughout the world have not yet recognized the desirability of population-based approaches to the provision of health services. A notable exception is McMaster University in the province of Ontario, Canada, which has attempted since its inception to integrate the skills of population medicine with those of clinical practice. The Rockefeller Foundation in the United States has used this model to encourage the development of programs in many medical schools throughout the world.

Several Latin American countries are actively pursuing an agenda for COPC. In Cuba, an intensive effort to train family physicians has been underway since the mid-1980s. These new physicians live and work with an associated nurse in the community they serve, which is intended to be about six hundred to seven hundred people. The family doctors spend half their time in the community providing services in organizations such as schools, day care centers, and factories and collecting information on community health needs. These physicians are expected to document the frequency of health problems in their practices and plan their clinical work to address population needs (UNICEF, 1991). In Mexico, Costa Rica, Nicaragua, and several other countries, medical education is directed at training physicians with a community orientation; in some schools, work in the community is part of the curriculum during each year of training in medical school (Braveman and Mora, 1987).

Fortunately, several trends in the delivery of health services, even in the United States, will facilitate community-oriented approaches to primary care (Rogers, 1982). To begin with, the financial remuneration for practicing primary care is more likely to improve than that for the practice of subspecialty medicine at least in the United States (Hsiao et al., 1988). Second, the physician will no longer be the sole "captain of the ship." As patients survive longer and the burdens of morbidity increase in community settings, other health personnel will become increasingly central to the avoidance of morbidity and the maintenance of well-being. Third, heightened attention will be given to the need to make training programs for practitioners more relevant to the changing population needs. Fourth, realization of the need for more effective and more efficient health care will draw attention to COPC. COPC can help communities organize more efficiently to prevent disease and promote health and encourage more discriminating use of medical technologies. Last, the advent of increasingly better information processing resulting from highly efficient and high-capacity computers will greatly facilitate the management of data from a variety of sources and linkages of data from diverse sources.

Identifying Community Health Needs

Community health needs are notoriously difficult to specify precisely and completely. Part of the difficulty derives from the problems in defining *needs,* and part from the difficulties in ascertainment.

What is meant by a *need?* Are needs to be defined primarily by the occurrence of conditions that result in premature death, or is prevalence of the condition itself a sufficient cause for concern? Or should the disability associated with health problems determine the priorities? Or the extent to which conditions result in absence from work or school? Or, alternatively, should needs be defined by failure to implement preventive strategies, jeopardizing health in the community?

The definition of need will vary from time to time and from place to place, depending on society's values and the availability of data. In large measure, the definition is a matter of values: who values what for whom. Because some needs are easier to measure than others, the definition will also depend on what can be assessed at any given time or place.

In most industrialized countries, vital statistics systems provide basic information about the causes of death and various routinely ascertained health problems, such as low birth weight or reportable diseases. Vital statistics have been used for many years as a starting point for assessment of community health needs and continue to be useful. For example, when infant mortality rates are greater in one city than in another, an investigation into the reasons for the differences is often launched. Needs are therefore defined empirically, that is, by comparing one area with others.

One technique for measuring needs involves a list of *sentinel* conditions that should occur rarely, or at best infrequently, in the presence of adequate care. The method is a systematic application of differences observed from vital statistics (Rutstein et al., 1976). The list contains over one hundred conditions or diagnoses. For most of the conditions on the list, death is preventable with early and appropriate medical care. For others, either preventive or treatment strategies will prevent premature death from the condition. Death statistics or, in some cases, case registries or hospitalization statistics provide the means for identifying these sentinel conditions. The occurrence of a condition on the list should signal an investigation to determine if it occurred because of inadequate or inappropriate health services in the community. This method is a "normative" approach to defining health needs, because definitive standards are set for at least some aspects of ill health.

Another normative approach to the assessment of health needs is *goal-setting*. This method gained currency in the early 1980s with the publication of the Surgeon General's "Goals and Objectives for the Nation" (Public Health Service, 1980). These goals derived from committees of experts who selected important health problems they believed to be sensitive to change. Each community could adopt all or some of the goals and could target resources to meet them where the baseline frequencies suggested that a treatable problem was present. Many goals for the year 1990 were met on the national level. The degree of variability in attainment of the goals at local or even state levels has not been ascertained, although it is likely that many local communities adopted the goals or modified them.

Goals for the United States for the year 2000 were published in 1990 (Public Health Service, 1990). There are 298 specific objectives in 22 priority areas, divided into four main types: health promotion, health protection, preventive services, and surveillance. These objectives are adaptable for use at the community or state level. In small communities, where the frequency of events is low or estimates of frequency may be unstable, data may have to be aggregated over several years to be useful.

The empirical approach to defining health needs has also undergone refinement. Computerization of health records and standardized reporting of diagnoses from hospitalizations makes possible the detection of differences in rates of hospitalization in local communities. Known as the *small area variation* approach, this method provides information that suggests the existence of systematic differences in health needs among different communities. Small area variations can occur for several reasons, including true differences in morbidity because of different environmental exposures or variations in medical practice that result in different propensities to hospitalize for the same condition. Systematic differences in hospitalizations for conditions thought to be amenable to prevention or believed to be treatable in an early stage, making hospitalization unnecessary, suggest the existence of health needs that are not being adequately met in some communities. Billings et al. (1989) have applied this method to a study of hospitalizations in one large metropolitan area. Conditions thought to be "ambulatory-care sensitive" were chosen for investigation, and differences in hospitalization rates for these conditions were ascertained. Communities with comparatively high rates were designated as having ambulatory care services that were inadequate in identifying or dealing with the health problems in the community.

Any community with higher than expected rates of hospitalization for ambulatory-sensitive conditions should examine their ambulatory services to determine if they are adequate in providing needed care in the community. For example, would additional social services or visiting nurses help keep individuals under care at home when the severity of their illness alone does not require hospitalization? Are health services located so that needy individuals can reach them? Are the economic incentives for providing certain services perverse? Are the physicians' services of poor quality? Or is the excess hospitalization just a result of idiosyncratic differences in the propensity of local physicians to hospitalize patients?

Individual primary care practices can also engage in activities that identify and respond to community health needs. One such effort, the patient advisory council concept, has been used successfully in a fee-for-service family practice in Minnesota. The practice defined its patient population and invited all members to join the council and attend the four meetings held each year. In addition, ongoing activities of the council are directed at improving various aspects of the services. Decision-making in the practice is shared with the council. One committee of the council (services improvement committee) had helped the practice better understand the needs of the community. As a result, the practice has lower malpractice premiums than do others in the area. In addition, epidemiologic data show that the practice identifies psychosocial problems in the population at rates comparable to those ascertained from community studies, whereas the rates in other practices are much lower (Seifert and Seifert, 1982).

As the capacity to collect data from an increasing number and variety of sources increases, new possibilities will arise for identifying different types of community health needs. A standardized data set for ambulatory care which routinely reports all problems and diagnoses made in office and center-based practice will illuminate the existence of problems not generally causing death or hospitalization. New systems for eliciting and categorizing different types of disability, limitation of activity, and interference with social roles as a result of health problems will make possible the use of these types of information to assess community health needs in ways other than traditional morbidity and mortality statistics. Better and less expensive approaches to surveying populations outside of health facilities will provide yet another avenue for identifying new types of health needs and, in addition, will provide a more complete picture of health needs than that obtained from data on individuals who have already received services.

The development of technology for collecting and processing information will certainly facilitate achievement of the initial steps in COPC. Improved medical technology will expand the definition of *health needs;* as existing problems are solved, new challenges at another level of need will emerge. Community-oriented care may not be achieved everywhere, or to the same degree in all places. But the concept is now appropriate for consideration as the challenges of the twenty-first century approach.

References

Billings J and Hasselblad V. A Preliminary Study: Use of Small Area Analysis to Assess the Performance of the Outpatient Delivery System in New York City. Lyme N.H., The Codman Research Group, Inc., 1989.

Braveman P and Mora F. Training physicians for community-oriented primary care in Latin America: model programs in Mexico, Nicaragua, and Costa Rica. Am J Public Health 1987; 77:485–90.

Dutton D. Children's health care: the myth of equal access. In: Select Panel for the Promotion of Child Health, Better Health for Our Children: A National Strategy. US DHHS (PHS) Pub. No. 79-55071. U.S. Government Printing Office, Vol 4, 1981, pp 357–440.

Hart JT. A new type of general practitioner. Lancet 1983; 2:27–29.

Holtzman N. Rare diseases, common problems: recognition and management. Pediatrics 1978; 62:1056–60.

Hsiao W, Braun P, Dunn D, Becker E, DeNicola M, and Ketcham T. Results and policy implications of the resource-based relative-value study. N Engl J Med 1988; 319:881–88.

Institute of Medicine (IOM). Community Oriented Primary Care: A Practical Assessment. Vol 1. Washington, D.C. National Academy Press, 1984.

Kark SL. Community Oriented Primary Health Care. New York, Appleton-Century-Crofts, 1981.

Kark SL. From medicine in the community to community medicine. JAMA 1974; 228:1585–86.

Mant D and Anderson P. Community general practitioner. Lancet 1985; ii:1114–17.

Nutting P and Connor E. Community Oriented Primary Care: A Practical Assessment. Vol. II. Case Studies. Washington, D.C. National Academy Press, 1984.

Nutting P and Connor E. Community-oriented primary care: an integrated model for practice, research, and education. Am J Prev Med 1986; 2:140–47.

Pickles WN. Epidemiology in Country Practice. Baltimore, Williams and Wilkins, 1939.

Rogers D. Community-oriented primary care. JAMA 1982; 248:1622–25.

Rutstein D, Berenberg W, Chalmers T, Child C, Fishman A, and Perrin E. Measuring the quality of medical care: a clinical method. N Engl J Med 1976; 294:582–88.

Seifert M and Seifert M Jr. The patient advisory council concept. In: Connor E and Mullan F (eds). Community Oriented Primary Care. Washington, D.C., Conference Proceedings, Institute of Medicine, National Academy Press, 1982.

Starfield B, Simborg D, Horn S, and Yourtee S. Continuity and coordination in primary care: their achievement and utility. Medical Care 1976; 14(7):625–36.

UNICEF/UNFPA/OPS/OMS/MINSAP. Cuba's Family Doctor Programme. Havana, Cuba, March 12–16, 1991.

USDHHS (U.S. Department of Health and Human Services). Public Health Service. Promoting Health/Preventing Disease. Objectives for the Nation. Washington, D.C., Fall 1980.

USDHHS (U.S. Department of Health and Human Services). Public Health Service. Promoting Health/Preventing Disease: Year 2000 Objectives for the Nation. DHHS Publication No. (PHS)90-50212. Washington D.C., 1990.

White KL, Williams TF, and Greenberg BG. The ecology of medical care. N Engl J Med 1961; 265:885–92.

C

Primary Care Systems

14

Evaluation of Primary Care Programs

This chapter provides examples and a general framework for evaluating the attainment of primary care objectives in the context of an operating program. It describes four different approaches: from a practice perspective, from a professional perspective, from a population perspective, and from an educational perspective. The chapter concludes with a summary of the attributes of primary care that can be evaluated as well as mechanisms for conducting the evaluations.

Evaluation from a Practice Perspective

This section describes two examples of evaluations of primary care in practice settings. One setting was a clinic for adults and the other a clinic for children.

Table 14.1 shows the characteristics of an evaluation of a primary care clinic for adults. The practitioners in the clinic established, a priori, a level of performance they considered appropriate. The performance of the primary care clinic was then assessed against these *absolute* standards.

One aspect of the study involved the attainment of coordination of care (Barker et al., 1989). Patients who were keeping an appointment for a follow-up visit were chosen for study if the follow-up visit took place within six months of the prior visit. Before the follow-up visit, records were reviewed to obtain all items of information from the prior visit. After the visit the records were again reviewed to ascertain recognition of the information.

The investigation showed the following:

- Recognition of problems identified at the index visit fell just short of the criterion level of 75 percent, but recognition was consistently better when the problem was on a problem list.
- Recognition of therapies that had been prescribed met the preset standards, but was better for major drugs (91 percent) and for minor drugs (70 percent) than for nondrug therapies (38 percent).
- Recognition of all tests scheduled at the prior visit (59 percent) did not reach the standards, both for tests with abnormal results and for those with normal results.
- Recognition was also inadequate for intervening visits, both scheduled and unscheduled.

As a result of the evaluation, the clinic directors altered the encounter form to require the separate recording of all nondrug therapies as well as intervening visits.

Another evaluation in the same clinic examined the adequacy of record keeping (Kern et al., 1990). Sixteen items were used to assess general record-keeping practices, as follows: legibility, past medical history, presence of a problem list on the front sheet of the patient's medical record, completeness of the problem list, allergy history on front sheet, smoking and alcohol history on the front sheet, completeness of the social history on the front sheet, clarity of recording of functional status, clarity of the medication regimen, recording of compliance, problem orientation of the visit notes, visit notes in SOAP format (see Chapter 10), use of a flow sheet for active problems, adequacy of care for active problems, documentation of patient education, and use of a flow sheet for preventive care.

Preventive care was assessed by the recording of ten elements: diphtheria/ tetanus vaccine within ten years, pneumococcal vaccine in high-risk patients, yearly influenza vaccine in high-risk patients, rectal/prostate examination every two years in patients age forty-five and above, stool occult blood examination yearly at ages forty-five and above, cervical cancer screening at least every three years, physician breast exam yearly for patients aged forty and older, breast self-exam taught once at age thirty-five or above, one serologic test for syphilis, and one tuberculin test. The chart audit took place for six consecutive years and adequacy of performance was defined as recording of indicated procedures in at least 75 percent of the records; performance of less than fifty percent was considered a major inadequacy. Only three of the twenty-six items reached the standard of 75 percent; fifteen had major deficiencies.

Improvements occurred over the six-year period in both categories of audit (general record keeping and preventive care documentation). The items that showed the least improvement were documentation of functional status, patient compliance, patient education (all over 50 percent initially) and documentation of a few items of preventive care.

In a study concerning children conducted at a tertiary medical center, a primary care pediatric practice was compared with office-based practices in the community, to provide a basis for judging the *relative* adequacy of performance. In the facility under study, physicians kept encounter logs and patients were asked four questions: Is this the child's first visit to the clinic? Did a doctor from someplace outside this hospital send you to this clinic? The last time this child had a regular checkup did he or she go someplace else or come here? The last time the child received medical care for a bad cold or flu did he or she go someplace else or come here?

Answers to these questions served as a basis for categorizing encounters as primary care, first encounter care, specialized care, or consultative care. A score was obtained and compared with scores obtained from office-based pediatric practices in the same community. The results of this evaluation indicated that the hospital clinic did as well as office-based physicians in the extent of longitudinality of care provided (Wilson et al., 1989).

In both studies, the clinic for adults and the one for children, the patient was the unit of analysis and the results were aggregated to characterize the clinical facility. The evaluations addressed three of the four unique features of primary care: coordination and comprehensiveness in the adult clinic and longitudinality in the pediatric clinic.

Evaluation from a Professional Perspective

The Baltimore City Medical Society initiated a cooperative venture involving state, regional, and city health planners and faculty and students in a school of public health to study the extent to which primary care resources were available in an entire U.S. city. The study had two purposes: to provide data for local planning of the primary care delivery system and to complement national data on measurement of the extent of primary care in office-based settings (Weiner et al., 1982).

Data were obtained from all physicians in the area and a stratified random sample were requested to complete an encounter log for each patient seen in one week.

Three methods were used to assess the extent of primary care in office-based practice. In the first method (the empirical method) an algorithm was used to distinguish among visits that were first contact, for consultation only, for ongoing nonreferral specialized services, and principal care. The last category included visits which were not for any of the other three purposes *and* when both the patient's last contact for a checkup and for care of a cold or the flu were to the doctor seen at the sample visit. The researchers developed an *Empirical Primary Care Index (EPCI)*, a weighted index of these four types of visits in which principal care was given a weight of 3, first encounter care was given a weight of 2, specialist care was given a weight of 1, and consultative care was given a weight of zero. Visits that did not fall into these categories were assigned a weight of 2.

The second method of assessment of the extent of primary care had four components, three of which were related to one of the unique attributes of primary care

TABLE 14.1.   Program Evaluation in a Primary Care Clinic for Adults
Aims and Mechanism of Evaluation

Aims	Mechanism
Longitudinality	Patient identification with a specific doctor
Comprehensiveness	Knowledge of patient's social profile Recognition of psychosocial problems Attitudes toward/knowledge of preventive and psychosocial needs
Coordination	Recognition of information from visits elsewhere
Record keeping	Documentation of medications and compliance Problem lists/POMR Preventive care
Communication skills	Review of videotapes
Quality of care	Control of hypertension

Source: Adapted from Barker et al., 1984.

while the fourth component concerned family-centeredness. The method is "normative" because the characteristics were judged against criteria that are assumed to be standards.

Comprehensiveness, it was assumed, would be greater if a certain defined set of services were available in the practice. These included physical examinations, immunizations, pelvic examinations, EKGs, blood hematocrit or hemoglobin determinations, and analyses requiring a microscope.

Accessibility of the practice was scored according to the availability of emergency appointments, willingness of the practitioner to make outside office calls (to home or emergency room), use of an answering service, and formal arrangements for after-hours coverage.

Longitudinal care was measured by the average duration of the patients' relationships with the particular physician seen and was controlled for the age of the patient.

The family-centeredness component was based on the percentage of patients in the practice who had at least one other immediate family member seen by the same physician.

The four separate scores were combined to develop the Normative Primary Care Index (NPCI).

In the third method of assessing the extent of primary care, Self-Assessed Primary Care Index (SAPCI), physicians were asked the question "What percentage of the visits at your main site would you estimate as being general medical care for patients for whom you maintain ongoing responsibility?" Physicians with higher percentages were assumed to be providing more primary care than physicians with lower percentages.

The results showed the following:

- The percentage of principal care visits was higher for family physicians, general internists, and pediatricians than for all other physicians.
- The percentage of first encounter visits was lower for psychiatrists and higher for medical subspecialists, but did not differ greatly for the other types of physicians.
- The percentage of specialized care encounters was high for obstetrician gynecologists and medical subspecialists, low for family physicians and pediatricians, and intermediate for all others.
- The percentage of consultative care visits was low for pediatricians and family physicians, and high for psychiatrists, surgical subspecialists, general surgeons, and medical subspecialists, and intermediate for the others.
- The EPCI (Empirical Primary Care Index), derived from the preceding scores appropriately weighted, was high for family physicians, pediatricians, and general internists, low for psychiatrists and surgical subspecialists, and intermediate for the others.
- The NPCI (Normative Primary Care Index) was highest for family physicians, pediatricians, and general internists, lowest for surgical subspecialists, and intermediate for the rest.
- The SAPCI (Self-Assessed Primary Care Index) was consistent with the results from the other two methods in indicating that internists, family physicians and pediatricians practice more primary care than any of the other specialists.

The findings lend credence to the usefulness of the measures in distinguishing achievement of the various characteristics of primary care in a population of practices.

Evaluation from a Population Perspective

Most assessments of the attainment of primary care are conducted from the viewpoint of the practice. Only a few evaluations consider the extent of primary care received by populations even though methods can assess attainment of both the structural and process feature unique to primary care from a population perspective (see Chapters 3–6).

One aspect of the Baltimore practice study described earlier involved an assessment of the availability and utilization of primary care services in the different areas of the city. Office-based visit rates were calculated by age, sex, race, and residency of patients within the areas. The measure of physician availability was visits per person. The study revealed the following:

- The residential area with the highest availability of physicians had 4.3 times the average number of visits as the area with the lowest availability.
- The percentage of the total ambulatory visits made to private offices varied from 41.6 to 96.5; overall, residents obtained only 37.5 percent of their office-based care within their home districts.
- The percentage of primary care visits to the four primary care specialities within their home districts was slightly higher (43.2 percent) than the percentage of all visits made within the home district.
- The use of office-based services varied markedly by race of the population: black residents made only 59 percent of the number of office-based visits as white residents.

Using U.S. data as a standard, Baltimore residents had an adequate level of utilization of specialty services but a much lower level of primary care visits than the national average. In five of the eight areas of the city, average family income was more closely related to use of services than availability of resources, whereas in the three other areas both income and availability of resources were equally associated with use. This implies that the presence of primary care resources in an area is a less important factor than the average income of its residents. Residents in high-income areas, compared with those in lower-income areas, had relatively high utilization, regardless of the availability of care in their home areas. In areas of low family income, utilization was low even though office-based primary care facilities were sometimes available.

A means of assessing the usefulness of population-based measures to describe some aspects of primary care was provided by two unique and linked data sets in Manitoba, Canada. A study analyzed data from linked survey and claims data from all physician visits (Black, 1990). Two measures of primary care were developed: a measure of longitudinality and a measure of comprehensiveness. The measure of longitudinality was created by applying the continuity of care measure (COC; see

Chapter 4) to the first visit in each episode of care for an illness. The measure was validated by testing it against what was reported by survey respondents' as their "regular source of care"; individuals who reported that they had a regular source of care had higher longitudinality scores. Comprehensiveness was measured by determining whether the individual received an immunization for influenza. Attempts were also made to assess both first contact care and coordination of care; although both methods were judged as promising, they did not attain standards for criterion validity.

The study indicated the following:

- Individuals who had been referred for consultation to another physician had lower scores for longitudinality, while those who received a greater proportion of services from primary care providers had higher longitudinality scores.
- Comprehensiveness was significantly related to the volume of primary care services individuals received.
- Among individuals who had referral visits, those with a greater proportion of visits from a primary care provider had greater comprehensiveness.
- In fact, the volume of primary care services was a better predictor of comprehensiveness than volume of total ambulatory visits.

This study demonstrates the potential for using data from claims forms to assess the attainment of at least two characteristics of primary care (longitudinality and comprehensiveness) on a population basis rather than from a practice perspective.

Evaluation from an Educational Perspective

An evaluation approach from an educational perspective uses the facility as the unit of analysis rather than the patient or the population as did the previously described evaluations.

The study judged the extent to which a grant program funded by the federal government succeeded in enhancing the training of primary care physicians. One aspect of the evaluation entailed a detailed mail survey of all training programs in pediatrics and in internal medicine and dealt with the measurement of primary care. All program directors were asked whether they currently have a primary care training program or had one in the past. A primary care training program was defined as one especially designed to train physicians for careers as generalists. In contrast, a traditional program was defined as one designed to provide training for physicians who will pursue careers delivering subspecialty medicine exclusively or combined with generalist care. It further defined a primary care training program as one that devotes more time to ambulatory care training than a traditional one, and this training emphasizes "continuity care" practice and training in settings in which residents provide continuous (not episodic) and comprehensive (not specialized) care (Noble et al., in press).

Separate analyses compared traditional programs with primary care programs on continuity scores, teaching comprehensiveness, and rotation comprehensiveness (Table 14.2). Federally funded primary care programs were compared with non-

TABLE 14.2. Program Comparisons:
Traditional vs. Primary Care

Continuity score*
Teaching comprehensiveness*
Rotation comprehensiveness*
Summary score (*)

*Components of summary score.

federally funded primary care programs for the three features: percentage of time spent in a continuity setting, comprehensiveness of teaching, and comprehensiveness of rotations. In addition, there was a comparison of the extent to which other features were achieved (Table 14.3). These included the extent of use of community training sites, faculty–trainee ratio, faculty breadth, minority recruitment, extent of internal evaluation of the programs, and the attainment of features of primary care. The features particularly pertinent to primary care practice included (a) access (a component of first contact care), (b) size of the panel followed over the time of the residency (a proxy for longitudinality), (c) the range of services provided (the structural component of comprehensiveness), (d) continuity (the structural feature of coordination), and (e) mechanisms to improve the recognition of information about patients (the process feature of coordination). The following questions tapped these elements of primary care:

> When the continuity practice is closed, as in the case of weekends or evenings, do patients have access to a physician member of the practice by telephone? (access measure)
> What is the average panel size of the trainees in the continuity practice (total number of patients) at the end of the third year of training? (longitudinality measure)
> When the continuity practice is closed, how are patients handled if they need to be seen by a physician before the next scheduled practice day? Three choices were listed. (continuity measure)
> What involvement are the primary care trainees expected to have with their continuity practice patients during hospital admission of these patients? Three choices were provided. (continuity measure)

TABLE 14.3. Program Comparisons:
Nonfederally Funded vs. Federally Funded

Continuity score*
Teaching comprehensiveness*
Rotation comprehensiveness*
Primary care score: continuity site*
Community training sites
Faculty/trainee ratio*
Faculty breadth*
Minority recruitment
Evaluation
Summary score (*)

*Components of summary score.

Who is responsible for ensuring the routine provision of primary and secondary preventive measures for the patients of trainees, such as routine immunizations, screening, health maintenance assessments? (comprehensiveness measure)

Are residents ever expected to make home visits? (comprehensiveness measure)

Are statistics compiled to profile the health problems of the patients served by the continuity practice? (comprehensiveness measure)

What methods (if any) are used to ensure that the indicated primary and secondary preventive measures are undertaken for individual patients? Seven methods were listed. (coordination measure)

What mechanism is used in the continuity practice to make trainees aware of the results of recent laboratory test/procedures and referrals to consultants for their patients? Three choices were provided. (coordination measure)

(An additional question was designed to ascertain the extent to which trainees were taught to manage a spectrum of problems without referral. The question ascertained the percentage of patients with the selected problems and sought to determine whether trainees handle them without referral to another health professional within or outside the continuity practice. The problems were minor laceration, need for tympanocentesis, failure to thrive, and behavior problems in pediatrics. In internal medicine the problems included acute painful shoulder, rash, pelvic pain, and symptomatic depression. Unfortunately a typographical error on the questionnaire rendered the question useless.)

In both internal medicine and pediatrics programs large and statistically significant differences were found between traditional and primary care programs for percentage of time spent in a continuity site, teaching comprehensiveness, and rotation comprehensiveness, and, hence the summary score of the three (Table 14.4).

Differences were found between federally funded primary care training programs and other primary care training programs in some areas of care and no differences in others. Areas with no differences included teaching comprehen-

TABLE 14.4.   Differences between Primary
Care and Traditional Training Programs

Dimension	Traditional	Primary care
Results: Pediatrics		
Continuity score[a]	36.5	59.5
Teaching comp[a]	75.4	84.6
Rotation comp[a]	20.0	26.8
Summary score[a]	130.1	170.9
Results: Internal Medicine		
Continuity score[a]	28.8	57.3
Teaching comp[a]	67.6	81.0
Rotation comp[a]	13.9	32.5
Summary score[a]	109.5	167.9

[a]Indicates significance at the .05 level with a one-tail test.

TABLE 14.5. Difference between Primary Care Training Programs,
by Source of Funding

Dimension	Nonfederally Funded	Federally Funded
Results: Pediatrics		
Continuity score[a]	35.5	65.9
Primary care score[a]	61.1	68.4
Summary score[a]	296.5	344.5
Results: Internal Medicine		
Continuity score[a]	37.6	65.8
Primary care score[a]	50.6	61.2
Summary score[a]	283.9	330.3

[a]Indicates significance at the .05 level with a one-tail test.

siveness, rotation comprehensiveness, use of community training sites, faculty–trainee ratio, faculty breadth, minority recruitment, and efforts at internal evaluation of the program or of the trainees. Some differences in attributes were unique to primary care. The continuity score was much higher for federally funded programs, an expected result since the terms of the training grant required the programs to achieve a certain level of continuity. In addition, however, the federally funded programs in both internal medicine and pediatrics had significantly higher primary care scores than those of other primary care training programs, as indicated in Table 14.5. The summary scores, made up of the continuity index and the primary care score, were also significantly greater for the federally funded programs in both pediatrics and general medicine.

In a mail survey such as this one, responses to questionnaires may be distorted by intentional or nonintentional inflation of responses perceived to be desirable, a problem less likely to appear in a personal discussion and observation. Site visits made as a part of this study confirmed the results from the mail survey and, if anything, suggested that the questionnaire findings underestimated the differences between the federally funded and the nonfederally funded programs.

A Framework for Evaluation

The foregoing studies demonstrate how the framework for assessing and evaluating primary care can be applied at various levels and in a variety of settings, such as populations, primary care facilities, and primary care training programs. The particular characteristics that are assessed depend on the goals of the program. When these are specified in advance, the techniques serve to determine how well the programs are achieving their intended purposes.

Three types of characteristics are involved in evaluating primary care programs: (a) unique features of primary care, (b) essential but not unique features, and (c) derivative features. Most evaluations of primary care will do more than focus on the unique features of primary care (first contact, longitudinality, comprehensiveness,

TABLE 14.6.   Measuring the Attainment of Primary Care

Feature	Type of Information Needed	Source(s) of Information
The Unique Features		
First contact care	Accessibility of facility	Program design
	Access to care	Interview
	Use of facility as place of first contact	Interview/audit/claims forms
Longitudinality	Knowledge of patient and patient's social milieu	Interview/record audit
	Use of the regular source of care	Interview/record audit/ claims data bases
	Length of relationship with patients regardless of type of need for care	Interview/record audit
	Definition of eligible population	Age/sex register
Comprehensiveness	Spectrum of problems dealt with	Program design
	Primary and secondary preventive activities	Record/audit/claims forms
	Recognition and management of psychosocial situations	Interview/record audit/ program design (e.g., home visits)
Coordination of care	% of people managed without a referral	Clinical information systems
	Mechanisms for continuity	Program design (personnel; records)
	Recognition of information from prior visits	Record audit
	Referral/consultation visits (occurrence & results)	Program design/record audit
Essential but Not Unique Features		
Medical record format	Problem list	Record audit
	Completeness of the medical record	Record audit
Continuity of care	% seeing same practitioner on follow-up	Record audit
Practitioner-patient communication	Content/quality of interaction	Observation/recording
Derivative Features		
Family centered	Knowledge of family members	Interview/record audit
	Knowledge of health problems of family members	Interview/record audit
Community oriented	Knowledge of community health needs	Interview/statistics compiled
	Participation in community activities	Interview
	Community involvement in practice	Program design

and coordination) but also will incorporate their assessment with those of related characteristics. Table 14.6 summarizes the various characteristics, the type of information needed to assess them, and the sources of information required for the purpose. To provide more detail than the table gives, a brief description of each of the characteristics follows.

The Unique Features of Primary Care

Chapters 3 through 6 provide the theoretical basis for these characteristics of primary care and for ways of assessing them. The mechanics of assessment requires certain types of information and methods of obtaining them as well as specific indicators of the characteristics.

First Contact Care. Three areas of information are required for assessment. The first involves accessibility of the facilities, a factor that can be determined from the program design. Special characteristics of the evaluation involve (a) hours of availability, (b) accessibility to public transportation, (c) provision of care without requirements for payment in advance, (d) facilities for the handicapped, (e) after-hours arrangements, (f) ease of making appointments and waiting time to appointments, and (g) absence of language and other cultural barriers.

The second area of information concerns the patient's experiences with regard to access to care. This can be determined by interviewing patients and populations for their views of the temporal, spatial, organizational, or cultural characteristics, as noted in the preceding paragraph.

The third area of information concerns the actual use of the facility as a place of first contact. Such information is obtained by interviewing patients and auditing the medical record or claim forms to determine the place of visit for newly experienced problems or for health needs.

Longitudinality. Assessment of longitudinality demands a knowledge of the patients in the program and their social milieu. This can be learned by interviewing patients and physicians and by examining the record for important information about them. Special characteristics to be considered are social history and exposures, work history and exposures, housing, diet, health history, family history, and genetic profile. Special characteristics to be determined from patients are identification with the regular source of care and satisfaction with the long-term relationship.

The regular source of care can provide information to evaluate longitudinality. Interview of patients for regularity of use for nonreferred care will provide it. Specific areas concern the degree to which patients always use the primary care source for visits, including disease management, management of signs and symptoms, administrative purposes (need for certification of illness and health), test results, preventive care, need for and return from consultation or referral, and prescriptions for medications or other therapies.

A third type of information required for evaluation of longitudinality involves the length of relationship between a patient and a practitioner, apart from specific disease management. Interviewing patients and examining the record will supply this information.

Comprehensiveness of Care. Here again, three types of information are needed for assessment. The first involves the range of activities the facility is designed to handle. Such information can be determined by examining the facilities and available equipment and the rates of use of provided procedures. Specific areas for consideration include the ability to provide care for short-term, long-term, and recurring illnesses; health education; minor surgery; mental health; and referral to ancillary services.

A second type of information involves the performance of indicated primary and secondary preventive activities such as immunizations, health education, and indicated screening procedures. They can be ascertained from medical records and claims forms.

The third type of information involves recognition and management of psychosocial problems. Such information is available from the medical record, by interview of patient and physician, and from program design, for example, the ability to make a home visit when one seems advisable. Several techniques evaluate this characteristic, including determining the rate of recording of diagnoses of the major psychosocial conditions in each age group and whether it conforms with known rates in the population served; adequacy of recognition of psychosocial conditions in comparison with the results of standard screening inventories administered to patients; and use of home visits for suspected social determinants of illness (allergies, poor heating conditions, poor sanitation, hazardous substances and environments).

Coordination of Care. The first area to be examined is the mechanism for continuity. The program design and personnel records will supply this information. To be taken into consideration are continuity of practitioner or team, ready availability of records, and ease of retrievable information. Second, one should examine the record for recognition of information from prior visits. A third item of information concerns referrals for consultation or for ongoing care. This can be learned from the program design and record audit. Special characteristics are the use of an organized system for referral and for retrieval of information concerning the results of referral, and documented recognition of the results of referrals.

Essential But Not Unique Features of Primary Care

Several characteristics are important in primary care, although their importance is not limited to it; they are important in speciality care as well as in primary care. These include a responsibility to be organized and efficient in providing care; to understand, advise, and guide patients; or to provide advocacy for patients when it is needed. These characteristics are important at *all* levels of care.

In assessing the aspects of primary care that are essential but not unique to it, five areas require investigation: (a) adequacy of medical records, (b) continuity of personnel, (c) practitioner–patient communication, (d) quality of care, and (e) advocacy for patients. Each has its particular type of information source for information, and its method for evaluating the characteristic.

Adequate Medical Records. Every record should contain an updated problem list which includes new problems and deletes resolved ones. Information concerning the problem list can be obtained from the medical record or from a computerized printout.

The format of the encounter notes also yields relevant information, as does completeness of the medical records. Each facility may have its own requirements, such as updated medication and allergy lists, preventive care or laboratory flow charts, timely entry of the results of pertinent laboratory tests or referrals, or a problem-oriented format for the encounter notes. Evaluation of all of these requires record audits, although these may need to be supplemented by other information to judge whether the record is complete.

Continuity of Personnel. Continuity of personnel from one visit to the next can be learned from the medical record. The important factor is the percentage of patients who, on follow-up, see either the same practitioner or a team member.

Practitioner–Patient Communication. This characteristic concerns the content and quality of interaction between the patient and the practitioner. Such information can be obtained from audiotapes and videotapes or by interviewing patients and physicians. Aspects to be assessed are agreement on the patient's problems, joint understanding of procedures for diagnostic assessment and management strategies, and the patient's satisfaction with the physician's approach to understanding the patient's needs. Evaluation of practitioner–patient communication might also involve indication that practitioners allow patients to play an active role in their care by responding constructively to their questions and their concerns and by seriously considering their suggestions for alternative approaches.

Quality of Care. Information concerning the processes and outcome of care can be found in the record, by observation, by interviewing patients, and by simulating patients. Aspects of quality include performance of appropriate generic and disease-specific processes of care including diagnostic and therapeutic procedures as well as reassessment, monitoring, and surveillance for adequacy of biologic, psychologic, and social responses and functional status.

Advocacy for Patients. Assessment of advocacy by a program is based on the degree of awareness and the extent of utilization of the many health, social, occupational, and fiscal agencies that have an impact on health and health services. Interviews of patients and physicians can determine their awareness of these agencies and of the resources they can provide. Aspects that characterize advocacy are securing needed improvements in housing, work conditions, neighborhood safety and sanitation, and financial resources related to the improvement and maintenance of health.

The Derivative Features of Primary Care

A high level of achievement of the unique and essential qualities of primary care results in two additional aspects called *derivative features*. These are *family centeredness* and *community orientation*. Both features extend beyond what is narrowly conceived as medical care.

Family centeredness results when the achievement of longitudinality provides a basis for considering patients within their milieux, the assessment of needs for comprehensive care considers the family context and its exposure to health threats, and the challenge of coordination of care confronts the limitations of family resources.

Family centeredness is evaluated by interviewing family members and comparing the information obtained with either interviews of physicians or information from the medical record. Areas of particular interest include awareness of intra-family communication and support, and appreciation of family resources including its educational levels and financial means.

A second aspect of family centeredness requires a knowledge of the health problems of family members. This information can also be obtained by interviewing patients and physicians and by examining the record. Knowledge about the patterns of illness in facilities can shed light on possible mechanisms of etiology and response to therapy.

Community orientation, the second derivative features, results from a high degree of comprehensiveness of general care. All health-related needs of patients occur in a social context; recognition of these needs often requires knowledge of the social context. Patients may not realize that they need health services, either because they lack knowledge about the importance of a preventive strategy or because they do not realize a problem has a medical basis or might be amenable to medical interventions. Understanding of the distribution of health characteristics of the community and of the resources available in the community provides a better basis for assessing health needs than interactions with patients or with the families of patients alone.

Evaluation of community orientation requires three types of information. The first is a knowledge of the health needs of the community, obtainable by interviewing physicians and comparing their responses with community statistics. Areas that might be examined are morbidity and mortality statistics, health surveys including disability levels as well as illnesses, and school or work absence rates.

A second type of required information is the extent to which practitioners become involved in community affairs, a facet ascertainable by interviews. Specific characteristics include knowledge about available social networks and support systems including all recreational, religious, political, and philanthropic resources.

A third type of information involves the extent to which the practitioner involves the patient community in practice-related issues, such as provision for a community or patient advisory council or periodic surveys of patient satisfaction and suggestions for improvement.

References

Barker L, Starfield B, Gross R, Kern D, Levine D, and Fishelman P. Recognition of information and coordination of ambulatory care by medical residents. Medical Care 1989; 27:558–62.

Black C. Population-based measurement of primary care to study variations in care received by the elderly. Unpublished dissertation, Baltimore, The Johns Hopkins University, 1990.

Kern D. Harris W, Boekeloo B, Barker LR, and Hogeland P. Use of an outpatient medical record audit to achieve education objectives: changes in residents' performances over six years. J Gen Intern Med 1990; 5(3):218–24.

Noble J, Friedman R, Starfield B, Ash A, and Black C. Career differences between primary care and traditional trainees in Internal Medicine and Pediatrics. Ann Int Med 1992; 116(6):482–87.

Weiner J, Kassel L, Baker T, and Lane B. Baltimore City Primary Care Study: the role of the office-based physician. Maryland State Med J 1982; 31:48–52.

Wilson M, Weiner J, Bender J, Bergstrom S, and Starfield B. Does a residents' continuity clinic provide primary care? AJDC 1989; 143:809–12.

15

Cross-national Comparisons of Primary Care

Health systems vary from country to country and even within countries. Chapter 8 (Organization and Financing of Care in the United States) indicated the wide variety of arrangements within that country and showed the impact these differences have on the characteristics of primary care. The first part of this chapter describes the nature of the most common form of primary care in ten industrialized nations.* The second part presents information about several aspects of the health of people in each of the same countries. This is followed by a discussion of the relationship between primary care and health levels in the selected countries and the implication of the relationships for policy concerning primary care within the context of health services systems.

Characteristics of Primary Care in Ten Countries

Comparable information on seventeen important characteristics of primary care was available for the ten countries. Eleven of the characteristics concern the health system and six concern those characteristics of practice that are most relevant to primary care. System characteristics include the type of system, the type of physician who provides primary care, financial access to health services, extent of restriction on access to specialized care, percentage of physicians who are specialists, income of primary care physicians relative to specialists, geographic organization of primary care services, most common site of primary care, reimbursement for services by primary care providers and by specialists, type of physician caring for hospitalized patients, and whether or not specialists are restricted to hospital practice.

Six characteristics concern the nature of practices including the four unique features of primary care (first contact, longitudinality, comprehensiveness, and coordination), and the two derivative features (family orientation and community orientation).

The following section describes each country according to each of these seventeen characteristics.

*The information is derived largely from five sources: Stephen (1979), The Swedish Health Services in the 1990s (1981), Schroeder (1984), Fry and Hasler (1986), and Weiner (1987).

Descriptions of the Ten Countries

Australia

Type of system: Unregulated primary care
By whom? Family (general) practitioner
Financial access to care: National health insurance
Percentage of active physicians who are specialists: 56%
Ratio average income primary care physicians to specialists: Approximately 1:2
Primary care organized by defined geographic area? In practice, not by formal designation
Most common site of primary care practice: Physicians' offices
Reimbursement of primary care practitioners: Fee-for-service
Reimbursement of specialists: Fee-for-service (in office practice)
GPs care for patients in hospital? No
Specialists restricted to hospital? No

First contact: Access to specialist by referral from primary care
Longitudinality of physician: Yes, patients generally relate to a single physician over time
    but no formal enrollment
Comprehensiveness: Moderate. Little adult prevention. Childhood immunization from
    government health centers
Coordination: Poor between general practitioners and specialists
Family centeredness: Yes
Community orientation: Moderate to poor

Belgium

Type of system: Unregulated primary care
By whom? Family (general) practitioner
Financial access to care: National health insurance
Percentage of active physicians who are specialists: 47%
Ratio average income primary care physicians to specialists: not available
Primary care organized by defined geographic area? No
Most common site of primary care practice: Physicians' offices
Reimbursement of primary care providers: Fee-for-service including copay by patients
Reimbursement of specialists: Fee-for service, including copay for consultations
GPs care for patients in hospital? No
Specialists restricted to hospitals? No

First contact: Access to specialist by self-referral or referral from primary care
Longitudinality of physician: Poor
Comprehensiveness: Poor
Coordination: Poor
Family centeredness: Yes
Community orientation: Poor

Canada

Type of system: Partially regulated private (financial incentives for practicing in
    underserved areas)
By whom? Family (general) practitioners
Financial access to care: National health insurance, operated through provinces
Percentage of active physicians who are specialists: 48%

Ratio average income primary care physicians to specialists: approximately 2:3 (higher in rural areas)
Primary care organized by defined geographic area? No
Most common site of primary care practice: Physicians' offices
Reimbursement of primary care physicians: Fee for service (negotiated rates)
Reimbursement of specialists: Fee for service (negotiated rates)
GPs care for patients in hospital? Yes, but declining in frequency in urban areas
Specialists restricted to hospitals? No

First contact: Access to specialist by self referral or referral from primary care; Specialists paid at a higher rate if patient referred.
Longitudinality of physician: Variable
Comprehensiveness: Good. All outpatient and inpatient care including preventive care, and psychosocial problems seen by prior appointment; out patient drugs, eye and dental care depends on province.
Coordination: Poor. Little transfer of information between primary care and specialty physicians.
Family centeredness: Generally good.
Community orientation: Variable

Denmark

Type of system: Regulated private primary care
By whom? Family (general) practitioners
Financial access to care: National health insurance
Percentage of active physicians who are specialists: 65%
Ratio average income primary care physicians to specialists: 1:1
Primary care organized by defined geographic area? No, but each GP has a defined list
Most common site of primary care practice: Physicians' offices
Reimbursement of primary care physicians: Capitation; additional fee-for-services for encouraged procedures
Reimbursement of specialists: Salary
GPs care for patients in hospital? No
Specialists restricted to hospitals? Most (not ENT or GYN)

First contact: Access to specialist by referral from primary care except for otolaryngologists, ophthalmologists, and podiatrists
Longitudinality of physician: Yes, patients register with a single physician
Comprehensiveness: Moderate range; excellent preventive care for children; little emphasis on prevention for adults
Coordination: Poor coordination with specialist/inpatient care; moderate coordination with long-term services
Family centeredness: Yes
Community orientation: Moderate

Finland

Type of system: Public, center-based
By whom? General practitioners
Financial access to care: National health insurance
Percentage of active physicians who are specialists: 63%
Ratio average income primary care physicians to specialists: 1:1
Primary care organized by defined geographic area? Yes

Most common site of primary care practice: Health centers
Reimbursement of primary care physicians: Salary
Reimbursement of specialists: Fee-for-service
GPs care for patients in hospital? No
Specialists restricted to hospitals? No

First contact: Access to specialist by referral from primary care
Longitudinality of physician: Moderate
Comprehensiveness: Extensive range; excellent preventive care for children; poor
     preventive care for adults
Coordination: Poor coordination with specialists/inpatient care; good coordination with
     long-term care
Family centeredness: Moderate
Community orientation: High

Germany (Federal Republic)

Type of system: Unregulated private (legal prerequisites for control are not enforced)
By whom? Family (general) practitioner
Financial access to care? Mandatory health insurance through employers for individuals
     under threshold income level
Percentage of active physicians who are specialists: 46%*
Ratio average income of primary care physicians to specialists: not available
Primary care organized by defined geographic area? No
Most common site of primary care practice: Physicians' offices
Reimbursement of primary care physicians: Fee-for-service
Reimbursement of specialists: Fee-for-service (salary in hospital)
GPs care for patients in hospital? They may
Specialists restricted to hospitals? No (no access to hospital practice unless on hospital
     staff)

First contact: Access to specialists by self-referral or referral from primary care**
Longitudinality of physician: Poor
Comprehensiveness: Coverage of medical, dental, ophthalmologic services. Variable
     preventive care rendered.
Coordination: Poor
Family centeredness: Poor
Community orientation: Poor

Netherlands

Type of system: Regulated private primary care
By whom? Family (general) practitioners
Financial access to care? National health insurance (70% governmental; 30% private)
Percentage of active physicians who are specialists: 65%

---

*Low estimate as it represents the percentage of ambulatory physicians who are specialists. More recent data (1991) indicates that 79% of physicians are specialists; the comparable figure for the mid 1980s was probably slightly lower.

**A restriction on use of specialists was imposed in the late 1980s to discourage unnecessary use. Under this restriction, patients must obtain a referral certificate for visits to a specialist but it does not always prevent patients from seeking care with a specialist directly. Also, some specialists serve as primary care physicians for their patients (Iglehart, 1991).

Ratio average income primary care physicians to specialists: Well below 1
Primary care organized by defined geographic area? Yes
Most common site of primary care practice: Physicians' offices
Reimbursement of primary care physicians: Capitation (70% of population) or fee-for-service (30% of population)
Reimbursement of specialists: Fee-for-service
GPs care for patients in hospital? No
Specialists restricted to hospitals? Yes

First contact: Access to specialists by referral from primary care
Longitudinality of physician: Yes, patients register with a single physician
Comprehensiveness: General immunizations done in public sector. Limited laboratory services in office practices. Preventive services to adults poor
Coordination: Specialists required to send letters to general practitioners
Family centeredness: Yes
Community orientation: Moderate

Sweden

Type of system: Public hospital, evolving to public local health centers
By Whom? Specialist/physicians-in-training in polyclinics; general practitioners in health centers
Financial access to care: National health service (tax-financed)
Percentage of active physicians who are specialists: 77%
Ratio average income of primary care physicians to specialists: greater than 1.00
Primary care organized by defined geographic area? Yes for health centers; no for hospitals
Most common site of primary care practice: Hospitals ("polyclinics") 50% and health centers 50%, with intent to increase health centers and decrease polyclinics
Reimbursement of primary care physicians: Salary
Reimbursement of specialists: salary
GPs care for patients in hospital? No
Specialists restricted to hospitals? Yes (some health centers have some specialists)

First contact: Access to specialists by referral from primary care and self-referral
Longitudinality of physician: Moderate
Comprehensiveness: Good range of services; excellent preventive care for pregnant women and children; poor preventive care for adults
Coordination: Poor for specialists/good for long-term care
Family centeredness? Moderate
Community orientation: Good

United Kingdom (England & Wales)

Type of system: Regulated private
By whom? Family (general) practitioners
Financial access to care: National health insurance (system)
Percentage of active physicians who are specialists: 57%
Ratio average income primary care physicians to specialists: 1:1
Primary care organized by defined geographic area? Yes
Most common site of primary care practice: Physicians' offices
Reimbursement of primary care physicians: Capitation, salary, fee-for-service for designated procedures; designated amount for overhead expenses

Reimbursement of specialists: Salary
GPs care for patients in hospital? No
Specialists restricted to hospitals? Yes

First contact: Access to specialists by referral from primary care
Longitudinality of physician: Yes, patients register with a single physician
Comprehensiveness: Moderate range; good preventive care for children; poor preventive
    care for adults
Coordination: Poor coordination with specialists/inpatient services; moderate coordination
    with long-term care services
Family centeredness: Yes
Community orientation: Moderate

United States (Fee-for-service private practice)

Type of system: Unregulated private
By whom? Family practitioners, general internists, pediatricians, and sometimes other
    specialists
Financial access: By private arrangement, through employers, or through governmental
    programs (the elderly; the very poor)
Percentage of active physicians who are specialists: 87% (66% if general internists and
    pediatricians are included as primary care physicians)
Ratio average income of primary care physicians to specialists: .50–.70 depending on
    type of primary care physicians and type of specialist
Primary care organized by defined geographic area? No
Most common site of primary care practice: Physicians' offices
Reimbursement of primary care physicians: Fee-for-service
Reimbursement of specialists: Fee-for-service
GPs care for patients in hospital? Yes
Specialists restricted to hospitals? No

First contact: Access to specialists by self-referral and referral from physicians
Longitudinality of physician: Variable
Comprehensiveness: Variable range; variable extent of preventive care for children;
    variable extent of preventive care for adults
Coordination: Variable coordination with specialist and inpatient services; poor
    coordination with long-term care
Family centeredness: Good for family physicians; no for others
Community orientation: Poor

To rate the "primary care-ness" of each country, each characteristic was assigned a score from zero (connoting the absence or poor development of the characteristic) to 2 (connoting a high level of development of the characteristic). A score of
1 was assigned if there was moderate development of the characteristic. Table 15.1
describes the criteria for scoring the health system characteristics and the characteristics for scoring the practice characteristics. The unweighted scores for each
country were averaged to derive a "primary care score."

Table 15.2 summarizes the characteristics rated for each country and the respective primary care scores, which ranged from .2 (for the United States) to 1.7 (for the
United Kingdom).

The scores for each country reflect the characteristics of the most common form

TABLE 15.1.

Criteria for Rating of Health System Characteristics Related to Primary Care

Type of system. Regulated primary care or public health centers are considered to be the highest commitment to primary care. Regulated primary care implies that national policies influence the location of physician practice so that they are distributed throughout the population rather then concentrated in certain geographic areas. Public health centers are also assumed to represent the equitable distribution of physician resources. Intermediate scores connote systems where incentives for equitable distribution are present and moderately effective.

Type of primary care practitioner. Generalists (family or general practitioners) are the prototypical primary care physicians because the nature of their training is exclusively devoted to primary care practice. General pediatricians and general internists are considered 'intermediate' primary care practitioners because their training has a major subspecialty focus. Other specialists are not considered primary care physicians because their training is focused on subspecialty issues.

Financial access to care. Universal government-sponsored national health insurance or a national health entitlement is considered most conducive to access to primary care services. National health insurance sponsored by nongovernmental agencies is considered intermediate because of the absence of uniform benefits. Absence of national health insurance is not considered conducive to access to primary care.

Percentage active physicians who are specialists. A value below 50% is considered indicative of an orientation toward primary care. Values of 50–75% are considered intermediate and values above 75% are considered to indicate a specialty-oriented system.

Salary of primary care physicians relative to specialists. A high ratio (0.9 : 1 or above) of average salary of primary care physicians to specialty physicians is considered an incentive toward primary care. A low ratio (0.8 : 1 or less) is considered an incentive toward a specialty oriented system. Ratios between .8 and .9 are considered intermediate.

System characteristics not scored for primary care are where care is provided (since there is not evidence that one type of site is better than another), the type of reimbursement of generalists and of specialists (since the impact of type of reimbursement on incentives for primary care practice is not well known), whether or not generalists care for patients in hospitals (since there is little evidence on the impact of this feature of a health services system), and whether or not specialists are restricted to hospitals (since consultations with primary care physicians might be enhanced by limited specialty practice in the community). Even though the assignment of primary care services to a defined geographic area is considered conducive to community orientation and hence potentially pursuant to high level primary care, no points are assigned since community orientation is assessed directly.

Criteria for Rating Practice Characteristics Related to Primary Care

First contact. First contact implies that decisions about the need for specialty services are made after consulting the primary care physician. Requirements for access to specialists via referral from primary care are considered most consistent with the first-contact aspect of primary care. The ability of patient to self-refer to specialists is considered conducive to a specialty-oriented health system. Where there are incentives to reduce direct access to specialists but no requirement for a referral, an intermediate score is assigned.

Longitudinality. Longitudinality connotes the extent of relationship with a practitioner or facility over time that is not based upon the presence of specific types of diagnoses or health problems. Highest ratings are given where the relationship is based upon enrollment with a source of primary care, with the intent that all nonreferred or nondelegated care will be provided by the practitioner. Lowest rates are given where there is not an implicit or explicit relationship over time and intermediate scores are assigned where this relationship exists by default rather than intent.

(continued)

TABLE 15.1.   (*Continued*)

Criteria for Rating Practice Characteristics Related to Primary Care

Comprehensiveness. The extent to which a full range of services is either directly provided by the primary care physician or specifically arranged for elsewhere is the measure of comprehensiveness. Highest ratings are given to arrangements for the universal provision of extensive and uniform benefits and for preventive care. Intermediate ratings are given to arrangements for the provision of *either* extensive benefits or preventive care, or for concerted efforts to improve these for needy segments of the population. Low ratings are given when there is no policy regarding a minimum uniform set of benefits.

Coordination. Care is considered coordinated where there are formal guidelines for the transfer of information between primary care physicians and specialists. Where this is present for only certain aspects of care (such as long-term care), intermediate ratings are given. Low ratings reflect the general absence of guidelines for the transfer of information about patients.

Family centeredness. High ratings are given to explicit assumption of responsibility for family centered care. One point is assigned if variable or moderate. Low ratings are given if largely lacking.

Community orientation. High ratings are given where practitioners use community data in planning for services or for the identification of problems. Intermediate values are assigned where clinical data derived from analysis of data from the practice is used to identify priorities for care. Low ratings are given when there is little or no attempt to use data to plan or organize services.

---

of services only and not other systems that may provide more adequate primary care to certain segments of the population. Nevertheless, the distribution of scores for the "modal" form of organization indicates two general types of systems. In the first group are Belgium, the United States, and Federal Republic of Germany, all with scores below 1. The second group contains the other seven countries, all of which have scores above 1.

Levels of Health in the Ten Countries

It is often difficult to compare the health of people in different countries since there are few standard definitions or criteria for designating or counting many causes of death or illness. International collaborations have resulted in some types of data that are comparable in kind and quality across some countries. The health indicators presented in this section are generally recognized as useful for cross-national comparisons, at least in the ten selected countries.

Only indicators that are known to be influenced by health services are presented. For example, neonatal mortality is substantially influenced by both the availability of neonatal intensive care units and a system of prenatal care that provides entry into these units for infants at high risk of mortality and morbidity. Substantial reductions in postneonatal mortality follow from improvements in access to health services and, conversely, worsen when access is reduced. Other modifiable indicators are death rates and life expectancy. Access to health services can prevent illness as well as cure it, reduce its severity, or postpone death. To the extent that health practitioners can influence the life-styles of their patients and the early seeking of care for illness, the number of years of potential life lost is amenable to modification by health services. Another example of indicators amenable to modification by health services is the completeness of immunizations, which reflects not the only availability of vaccines but also the ability of a health system to deliver them when they are indicated. A final example of an indicator sensitive to health services interventions is low birth weight. Infants born weighing less than 2,500 grams are at substantially higher risk of morbidity throughout childhood as well as death in the first year of life (McCormick, 1985; McGauhey et al., 1991).

Table 15.3 presents death rates in the first month of life (neonatal mortality), in the remainder of the first year (postneonatal mortality), and for the first year as a whole (infant mortality) for each of the ten countries considered in this chapter. In 1986 (the latest year for which comparable data are available), neonatal mortality rates ranged from 3.95 in Sweden to 6.71 in the United States. Postneonatal mortality rates ranged from 1.84 in Finland to 4.48 in Belgium.

Table 15.4 presents age-adjusted death rates for the population as a whole, average life expectancy, and life expectancy at ages one, twenty, and sixty-five in the ten countries. It also presents years of potential life lost (YPLL), a measure of the component of mortality before age sixty-five that is considered preventable. The number of deaths in each disease category is multiplied by the difference between age sixty-five and the midpoint of each age category, so that deaths occurring at younger ages are more heavily weighted than those at older ages (CDC, 1986).

Tables 15.5 and 15.6, respectively, present death rates from injuries and from conditions other than injuries in four child age groups for the eight countries for which they were available.

Table 15.7 contains information on completed immunizations in the preschool age group in the countries for which data are available.

Table 15.8 presents the percentage of live-born infants weighing less than 2,500 grams at birth.

Another indicator of impact of a health services system is the population's satisfaction. In a survey conducted in 1988 and 1989 in the ten countries, people were asked their views about their own health system (Blendon et al., 1990). (The survey also included France, Italy, and Japan but did not include Belgium, Denmark, or Finland.) They were asked to choose one of the following options:

"On the whole the health care system works pretty well and only minor changes are necessary to make it work better."
"There are some good things in our health care system but fundamental changes are needed to make it work better."
"Our health care system has so much wrong with it that we need to completely rebuild it."

Three groups of countries were identified from this survey. In the first, the majority, or close to a majority, felt that only minor changes were needed. Among these countries were Canada and the Netherlands. In the second group, a majority or near-majority thought that fundamental changes or a complete rebuilding were needed, although about one-third thought only minor changes were necessary.

TABLE 15.2.   The Ten Countries

Country	Australia	Belgium	Canada	Denmark
Type of system	Unregulated primary care	Unregulated primary care	Partially regulated private (financial incentives for practicing in underserved areas)	Regulated private primary care
Type of primary care practitioner	Family (general) practitioner	Family (general) practitioner	Family (general) practitioners	Family (general) practitioners
Financial access to care	National health insurance	National health insurance	National health insurance, operated through provinces	National health insurance
% of active physicians who are specialists	56%	47%	48%	65%
Ratio of income of primary care physicians to specialists	Approximately 1 : 2	Not available	Approximately 2 : 3 (higher in rural areas)	1 : 1
First contact	Access to specialist by referral from primary care	Access to specialist by self-referral or referral from primary care	Access to specialist by self-referral or referral from primary care; Specialists paid at a higher rate if patient referred	Access to specialist by referral from primary care

Finland	Germany (F.R.)	Netherlands	Sweden	United Kingdom	United States
Public, center-based	Unregulated private (legal prerequisites for control are not enforced)	Regulated private primary care	Public hospital, evolving to public local health centers	Regulated private	Unregulated private
General practitioners	Family (general) practitioner	Family (general) practitioner	Specialist/physicians-in-training in polyclinics; general practitioners in health centers	Family (general) practitioners	Family practitioners, general internists, pediatricians, and sometimes other specialists
National health insurance	Mandatory health insurance through employers for individuals under threshold income level	National health insurance (70% governmental; 30% private)	National health service (tax-financed)	National health insurance (system)	By private arrangement, through employers, or through governmental programs (the elderly; the very poor)
63%	46%	65%	77%	37%	87% (66% if general internists and pediatricians are included as primary care physicians)
1:1	Not available	Well below 1	Greater than 1.00	1:1	.50–.70 depending on type of primary care physician and type of specialist
Access to specialist by referral from primary care	Access to specialists by self-referral or referral from primary care	Access to specialists by referral from primary care	Access to specialists by referral from primary care and self-referral	Access to specialists by referral from primary care	Access to specialists by self-referral and referral from physicians

(continued)

TABLE 15.2. *(Continued)*

Country	Australia	Belgium	Canada	Denmark
Longtudinality of physician	Yes, patients generally relate to a single physician over time but no formal enrollment	Poor	Variable	Yes, patients register with a single physician
Comprehensiveness	Moderate. Little adult prevention. Childhood immunization from government health centers	Poor	Good. All outpatient and inpatient care including preventive care, and psychosocial problems seen by prior appointment; outpatient drugs, eye and dental care depends on province.	Moderate range; excellent preventive care for children; little emphasis on prevention for adults
Coordination	Poor between general practitioners and specialists	Poor	Poor. Little transfer of information between primary care and specialty physicians	Poor coordination with specialist/inpatient care; moderate coordination with long-term services
Family centeredness	Yes	Yes	Generally good	Yes
Community orientation	Moderate to poor	Poor	Variable	Moderate
P.C. scores	1.1	.8	1.2	1.5

These countries include Germany (Federal Republic), Australia, Sweden, and the United Kingdom. In the United States, however, the vast majority (89 percent) thought fundamental changes or a complete rebuilding were required; only 10 percent thought only minor changes were needed.

Figure 15.1 presents the satisfaction–expense index derived from this survey for those countries covered in this chapter for which it is available. The index was derived by calculating the percentage who said the system needed only minor

Finland	Germany (F.R.)	Netherlands	Sweden	United Kingdom	United States
Moderate	Poor	Yes, patients register with a single physician	Moderate	Yes, patients register with a single physician	Variable
Extensive range: Excellent preventive care for children; poor preventive care for adults	Coverage of medical, dental, ophthalmologic services. Variable preventive care rendered	General immunizations done in public sector. Limited laboratory services in office practices. Preventive services to adults poor	Good range of services; excellent preventive care for pregnant women and children; poor preventive care for adults	Moderate range; good preventive care for children; poor preventive care for adults	Variable range; variable extent of preventive care for children; variable extent of preventive care for adults
Poor coordination with specialists/inpatient care; good coordination with long-term care	Poor	Specialists required to send letter to general practitioners	Poor for specialists/good for long-term care	Poor coordination with specialists/inpatient services; moderate coordination with long-term care services	Variable coordination with specialist and inpatient services; poor coordination with long-term care
Moderate	Poor	Yes	Moderate	Yes	Good for family physicians; no for others
High 1.5	Poor .5	Moderate 1.5	Good 1.2	Moderate 1.7	Poor .2

changes and dividing this by the percentage who said the system needed rebuilding, then dividing the resulting proportion by the per capita cost of the health care system per $1,000 spent. The graph shows the high satisfaction-to-cost ratio in the Netherlands and Canada, the intermediate ranking of Sweden, Germany (Federal Republic), Australia, and the United Kingdom, and the low ranking of the United States. Adjusting the denominator for differences in the age distribution of the population does not change the country ranking. For example, the adjusted index for

TABLE 15.3.   Neonatal and Postneonatal Mortality Rates,
10 Countries, 1986

	Neonatal	Postneonatal	Infant Mortality
Australia	5.43	3.42	8.85
Belgium	5.21	4.48	9.69
Canada	5.12	2.76	7.88
Denmark	5.10	3.09	8.19
Finland	4.01	1.84	5.85
Germany (Federal Republic)	4.77	3.77	8.54
Netherlands	4.80	2.96	7.76
Sweden	3.95	1.98	5.93
United Kingdom	5.28	4.27	9.55
United States[a]	6.71	3.64	10.35

Source: National Center for Health Statistics Data Bank. (Courtesy Robert Hartford and Sam Notzon.)

[a]The relatively high mortality rates for the United States as a whole are also found among the white population only, with a neonatal mortality rate of 5.81, a postneonatal mortality rate of 3.13, and a total infant mortality rate of 8.94

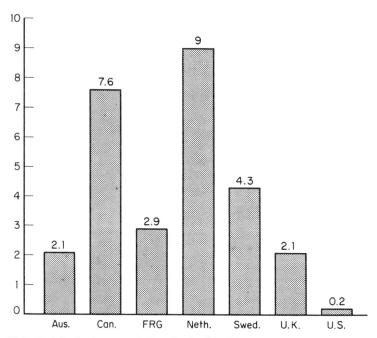

FIGURE 15.1. Satisfaction/expense index. Rank of health care systems in seven nations. *Source:* Health Letter 1990; 6(8):10.

TABLE 15.4. Age-Adjusted Death Rate, Life Expectancy, and Years of Potential Life Lost (YPLL)

| | Age-Adjusted Death Rates (1986–6) | Average Life Expectancy | | | | | | YPLL (1986–87) |
| | | At Age 1 (1980) | | At Age 20 (1980) | | At Age 65 (1986) | | |
		Females	Males	Females	Males	Females	Males	
Australia	774.9	77.9	70.9	59.4	52.6	18.9	14.9	4615.1
Belgium	889.4	76.1	69.3	57.6	51.4	17.8	13.6	5827.9
Canada	766.3	79.0	71.0	60.0	52.0	19.3	15.0	4547.9
Denmark	877.3	76.9	70.9	58.3	52.4	18.1	14.1	4910.2
Finland	881.1	77.1	68.7	58.4	50.3	17.7	13.5	4813.0
Germany (Federal Republic)	823.4	76.7	69.9	58.1	51.7	17.7	13.8	4507.1
Netherlands	788.2	79.1	72.2	60.5	53.8	18.9	14.1	3976.4
Sweden	752.4	78.5	72.4	59.8	53.7	19.0	14.9	3756.7
United Kingdom	857.6	76.6	70.6	58.0	52.2	17.9	13.9	4411.8
United States	828.4	77.4	69.7	58.9	51.5	18.6	14.7	5808.9

Source: OECD, 1986; CDC, 1990; NCHS, 1990.

TABLE 15.5.   Death Rates from Injuries in 4 Child Age Groups,
by Gender, 7 Countries, 1985

| | Death Rates from Injuries per 100,000 Population | | | | | | | |
| | Ages 1–4 | | Ages 5–9 | | Ages 10–14 | | Ages 15–19 | |
Country	Females	Males	Females	Males	Females	Males	Females	Males
Australia	18.4	24.3	7.7	15.5	5.0	13.1	21.3	67.1
Canada	12.3	19.3	8.6	14.7	8.3	16.6	21.1	55.7
Germany (Federal Republic)	11.3	15.1	8.7	12.2	4.5	9.2	15.1	48.4
Netherlands	7.0	12.9	3.0	9.4	4.8	7.2	9.0	27.6
Sweden	3.3	6.7	4.2	6.8	4.1	7.0	13.7	35.0
United Kingdom	6.9	12.2	4.4	8.4	5.3	13.9	9.9	35.6
United States	15.8	24.1	8.7	14.9	7.8	18.4	23.6	64.1

Source: NCHS, 1989.

TABLE 15.6.   Death Rates from Causes Other than Injuries in 4 Child Age Groups,
by Gender, 7 Countries, 1985

| | Death Rates from Causes Other Than Injuries per 100,000 Population | | | | | | | |
| | Ages 1–4 | | Ages 5–9 | | Ages 10–14 | | Ages 15–19 | |
Country	Females	Males	Females	Males	Females	Males	Females	Males
Australia	25.5	30.9	11.3	13.9	10.0	14.0	15.0	24.1
Canada	23.9	26.8	9.5	11.1	10.7	12.0	15.2	23.8
Germany (Federal Republic)	29.3	30.6	12.7	11.3	10.4	10.7	13.0	19.8
Netherlands	29.6	31.8	10.4	11.7	8.7	13.7	13.9	26.1
Sweden	24.6	23.3	5.5	12.4	9.9	10.2	10.0	14.9
United Kingdom	32.1	35.9	13.0	12.5	12.9	13.9	15.3	24.1
United States	25.9	30.9	11.4	12.2	10.6	12.1	14.9	20.4

Source: NCHS, 1989.

TABLE 15.7.   Completed Immunization Rates in Preschool Children[a],
1985–87, in 8 Countries

	DPT	Measles	Polio
Belgium	95.0	90.0	99.0
Denmark	94.7	82.0	100.0
Finland	90.0	81.0	95.0
Germany (Federal Republic)	95.0	50.0	95.0
Netherlands	96.9	92.8	96.0
Sweden	99.4	93.6	98.2
United Kingdom	87.0	76.0	87.0
	(pertussis 73.0)		
United States	64.9	60.8	55.3
(whites only)	(68.7)	(63.6)	(58.9)

Source: Williams and Miller, table B.10, p. 76.

[a] Age 3, except age 4 in United States.

TABLE 15.8.   Percentage of Live-Born
Infants Who Are of Low Birthweight,[a] 1983–84

Australia	5.8
Belgium	5.4
Canada	5.8
Denmark	5.7
Finland	3.7
Germany (Federal Republic)	5.5
Netherlands	4.0
Sweden	4.2
United Kingdom	6.7
United States	6.8

Source: World Health Organization, 1986.
[a]Birth weight less than 2,500 gram.

Canada, the Federal Republic of Germany, Sweden, the United Kingdom, and the United States are 6.6, 2.9, 4.4, 2.2, and .2, respectively, compared with unadjusted values of 7.6, 2.9, 4.3, 2.1, and .2, respectively.

Relationships Between Primary Care and Health Levels

There is a general concordance between the primary care score, the satisfaction index, and the health indicators (Table 15.9). For example, countries with primary care scores above 1 and with satisfaction/expense indices above 3 have generally better health indices. In contrast, the United States, Belgium, and Federal Republic of Germany rank low on all three. Countries where there are notable discrepancies between the rankings provide clues as to other factors that may be important in modifying the relationship between primary care and health status. For example, although Sweden has only a moderate primary care score (1.2), it has a very well-developed social service system that provides not only financial security throughout life but also specific arrangements to provide services that have an impact on health. Nurses regularly make home visits to provide preventive care for maternal and child health as well as indicated illness care. Long-term care services are highly developed, and in many places are directly attached to local health centers. In Canada (with a primary care score of 1.2), the social welfare system is highly developed: everyone is guaranteed a minimum income. A highly developed system of home visiting provides care for individuals with conditions that may benefit from it.

In contrast, the United Kingdom had the highest primary care score (1.7) but relatively poor performance on the health indicators. One possible reason for this discrepancy is the relatively low percentage of expenditures for both social services and education in the United Kingdom. The United Kingdom and the United States are the only two countries in the lowest third of the distribution both for the percentage of central government expenditures spent on housing, social security, and welfare and for education (Table 15.10). Table 15.10 also indicates that the percentage of central government expenditures spent on health services overall has

TABLE 15.9

| | Primary Care Score | Satisfaction/ Expense Index | Health Indicators[a] | |
			Top Third	Bottom Third
Australia	1.1	2.1	3	0
Belgium	0.8	NA[b]	0	9
Canada	1.2	7.6	5	0
Denmark	1.5	NA	0	3
Finland	1.5	NA	5	6
Germany (Federal Republic)	0.5	2.9	1	7
Netherlands	1.5	9.0	10	0
Sweden	1.2	4.3	10	0
United Kingdom	1.7	2.1	0	8
United States	0.2	0.2	1	7

[a]These columns present the number of indicators for which the country falls in the top (best) third of the distribution and the number for which it falls in bottom (worst) third. Sometimes there were three and sometimes four countries in the bottom third, depending on whether or not the countries had very similar values on the indicator. In the case of the top third, there were sometimes only two countries because they had values far better than the middle group which had values very close to each other. Although there were 31 indicators in all (including separate breakdown by age and sex), comparable information was available for all countries for only 12 of the indicators (and excluding death rates from injuries and natural cause by individual child age group, and immunization rates). Similar ratings are obtained, however, when the other indicators are added and used for comparisons among countries for which they were available.
[b]Not available.

little relationship to the level of health indicators. The three countries with the highest percentage spent on health services (United States, United Kingdom, and Federal Republic of Germany) are among the countries with the poorest performance on the indicators. Furthermore, a widening social disparity since 1950 between the upper and lower classes, which accelerated during the 1980s (Black, 1980; Whitehead, 1987), has been accompanied by increasing disparities in various health indicators. Access to primary care services may have little impact on health in a society in which as much as 60 percent of the disparity in health levels is

TABLE 15.10. Percentage of Central Government Expenditures
Spent in Different Sectors, Mid to Late 1980s

Country	Housing, Social Security, Welfare	Education	Health
Australia	28.6	7.0	9.6
Belgium	43.3	12.2	1.8
Canada	37.3	3.1	5.9
Denmark	41.1	9.0	1.3
Finland	36.1	13.9	10.6
Germany (Federal Republic)	49.4	0.6	18.2
Netherlands	39.6	11.9	10.9
Sweden	54.2	9.2	1.1
United Kingdom	30.9	2.2	13.6
United States	31.5	1.7	12.5

Source: World Bank, 1990.

attributable to social conditions of living, especially when increasing privatization of health services and decreased governmental support of the national health program are changing the character of the health system. In fact, the standing of the United Kingdom with regard to the health of its population has fallen relative to that of at least some other European nations (including Finland) during the period of these social changes (Townsend, 1990).

Finland provides an interesting example of a country whose emphasis has only recently shifted to primary care (Kekki, 1986). Neonatal and postneonatal mortality rates in Finland are among the best in the world whereas its age-adjusted death rates and life expectancy at older ages are among the poorest of the Western industrialized nations. If it is the recent refocusing of services on primary care that is contributing to the improvement in child health, it may by that adult health indices will improve as the population ages under the new system.

Implications for Primary Care of Cross-national Differences

If it were possible to relate differences in mortality and morbidity to differences in characteristics of health services systems, public policy decisions would be much easier. Unfortunately, it is unlikely that this will ever be possible, because health system structures influence care only through the processes of care. As earlier chapters in this book have demonstrated, it has been possible to link some processes of care to certain outcomes. If a link between the structures of the system and these processes could be demonstrated, inferences could be made about the relationships between the structures and the outcomes (health status). Unfortunately, the state of the art of health services research has not yet progressed to the point where these links have been adequately explored.

Moreover, health indicators reflect much more than the impact of heath services. As Chapter 2 indicated, health status is influenced by social and environmental factors as well as individual behaviors, many of which may not be amenable to change by health practitioners. Even if data and research methods were flawless, a perfect relationship between health services and health indicators would not be expected. But certain observations about the relationship between the health system features, as described earlier in this chapter, and the achievement of the goals of primary care provide a basis for hypotheses concerning the impact of differences in system characteristics on health status.

In his comparison of the health systems of five nations, Weiner (1987) paid particular attention to the relationship between system characteristics and the achievement of primary care, as assessed by access, comprehensiveness, coordination, and longitudinality, and by family centeredness and community orientation. His comparisons across the countries led him to conclude that several system characteristics were associated with greater achievement of the features of primary care. Longitudinality of care, the relationship over time between physician and patient, is facilitated by a capitation type of reimbursement; both Denmark and the United Kingdom are in this category. Center-based primary care facilitates community orientation; in both Finland and Sweden local councils of citizens have a major role in developing policies for the health centers. In many European countries, nurses play an important role in certain aspects of care; their involvement in home visiting as part of primary care and preschool and in-school prevention programs is responsible for high achievement of comprehensiveness of care for children. In Sweden and Finland, for example, about 40 and 25 percent, respectively, of ambulatory care visits are delivered by nurses, compared with fewer than 20 percent in most other countries. In the United Kingdom and in Denmark visiting nurses complement primary care services provided in offices or health centers.

Schroeder's analysis (1984) further supports the notion that certain structural features have a major impact on the attainment of primary care. In focusing particularly on personnel, he notes that merely increasing the number of physicians does not ensure an increased focus of primary care. The important goal is to improve the ratio of primary care physicians to specialists; this is likely to occur only if specialists are restricted to hospital practice and the number of training posts for specialists is limited. By comparing care in five countries, Schroeder tested six hypotheses concerning the impact of system characteristics on costs of care. He postulated that the following six conditions would result in the most costly systems.

- A high concentration of physicians and specialists.
- Fee-for-service payment of specialists and generalists.
- Patient self-referral to specialists.
- Physicians permitted to practice a specialty without specialty certification.
- A high dependence on specialists for primary care.
- Broad national health insurance.

Of the five countries examined in Schroeder's analysis, the United States meets the most conditions (five), lacking only national health insurance; Germany (Federal Republic) and Belgium meet four, lacking self-declared speciality practice and the prominent role of the specialist in primary care. The Netherlands meets three, differing from Belgium and Germany (Federal Republic) in not allowing self-referrals to specialists. The United Kingdom meets only one of the conditions (broad national health insurance). Comparisons of the percentage of gross national product spent on health care approximates predictions, with the exception that the percentage spent in the Netherlands exceeds that in Belgium despite the fact that the former meets fewer conditions. The United States both meets the most conditions and has the highest costs.

Other structural features of health systems can facilitate or hinder the achievement of effective care. Countries which have clear lines of responsibility for various aspects of health care outperform countries where the responsibility is either shared or diffused. For example, newborn screening in the United Kingdom and Ireland is much more effective than in the United States. In the former two countries, personnel and facilities have clearly defined roles and responsibilities, laboratories are centralized with good quality control, and follow-up care of young infants with suspected problems is ensured. In the United States, a pluralistic health system is associated with multiple lines of responsibility and difficulties in public accountability (Starfield and Holtzman, 1975). The Netherlands has a highly developed computerized information system that keeps track of children's immunization and notifies parents when an indicated immunization has not been received (Verbrugge, 1990).

These analyses of the relationships between various aspects of the structure and processes of care indicate that more systematic and more targeted studies will suggest directions for improved policymaking in health systems. The few descriptions of different health systems available suggest that some characteristics are likely to be more salient than others in producing a system oriented toward the principles of primary care. These are as follows:

1. A high societal commitment, as manifested by a government program of services that places a high value on primary care by paying for it and improving access to care.

2. Better use of nonphysician personnel, especially nurses who make home visits to perform designated functions, particularly those related to prevention of illness and disability to improve comprehensiveness of care.

3. Government regulation (if the modal practice form is in the private sector) to ensure appropriate distribution of services and balance between primary and specialty services.

4. Capitation reimbursement for primary care, with additional incentives to provide certain high-priority services, to improve both longitudinality and comprehensiveness of care.

5. Payment of specialists by salary rather than fee-for-services, so as to achieve a more appropriate and less costly allocation of resources.

6. Restricted access of patients to specialists to encourage first contact care by primary care providers and facilitate the development of mechanisms for improving coordination of services.

7. Designation of specific service areas so that practitioners have clear identification with and accountability to defined populations.

8. Organization of primary care services in health centers rather than in individual physicians' offices, to facilitate comprehensiveness of care and improve the likelihood of high-quality care.

9. Clear lines of responsibility and accountability for the performance of high-priority aspects of health care.

The influence of characteristics other than those listed here is less clear. In most industrialized nations primary care is provided by family physicians, usually called "general practitioners." The literature described in Chapter 9 suggests that internists and pediatricians differ from family practitioners in some specific ways, but evidence of the impact of these differences on health status or on costs of care is lacking.

The impact of limiting the provision of inpatient services to hospital-based specialists requires exploration. Practice in the United States is unusual in that it allows primary care physicians to care for their hospitalized patients. It would be useful to know if a designated in-hospital role for the primary care physician leads to more appropriate care or shorter lengths of hospital stay.

The effect of limiting the practice of specialists to hospitals is unclear. In some health systems in industrialized countries, specialists are an integral part of health centers; in other countries they are permitted to have their own offices outside of hospitals and health centers. Knowledge of the influence of these differences on coordination of services, costs, and quality of care would be useful.

In most industrialized countries, general practitioners comprise more than one-third of physicians, while in the United States general practitioners comprise only 13 percent of all active physicians (AMA, 1990). Adding pediatricians and internists to the number of primary care physicians brings the percentage to about 40 percent. In the United States, the proportion of physicians who are generalists has been falling steadily, while that of general internists and pediatricians has been rising; as a result, the proportion of physicians who are primary care physicians has remained steady. In some countries, increasing the proportion of generalists is public policy. For example, in Finland, there is a planned increase in primary health care physicians of 28 percent from 1982 to the year 2000; the number of specialists would increase only 13 percent during this period (Kekki, 1986). Health care costs are to grow by 77 percent during this time, but primary care costs will double.

Policymakers in all countries could benefit from information concerning the relative impact on health of systems oriented toward a growth in primary care, in contrast with those in which the growth of specialists is outpacing that of generalists. Although specific evidence on this issue is lacking, comparisons of health indicators across different countries provide clues that a primary care focus is associated with better health, at least as measured by certain indicators.

References

AMA (American Medical Association). Physician Characteristics and Distribution in the U.S. 1989 Ed. Department of Physician Data Services. Division of Survey and Data Resources, 1990.

Black D. Inequalities in Health. Report of a Research Working Group. Department of Health and Social Security (United Kingdom), 1980.

Blendon R, Leitman R, Morrison I, and Donelan K. Satisfaction with health systems in ten nations. Health Affairs 1990; 9:185–92.

CDC (Centers for Disease Control). Changes in Premature Mortality. U.S. 1983–1984. Morbidity and Mortality Weekly Report 1986; 35(2):29–31.

CDC (Centers for Disease Control). Mortality in developed countries. Morbidity and Mortality Weekly Report 1990; 39(13):205–09.

Fry J and Hasler J. Primary Health Care 2000. Edinburgh, Churchill Livingstone, 1986.

Health Letter 1990; 6(8):10.

Iglehart J. Germany's health care system. N Engl J Med 1991; 324:503–08.

Kekki P. Finland. In: Fry J and Hasler J. Primary Health Care 2000. Edinburgh, Churchill Livingstone, 1986.

Köhler L and Jakobsson G. Children's Health and Well-Being in the Nordic Countries. Oxford, MacKeith Press, 1987.

McCormick M. The contribution of low birth weight to infant mortality and childhood morbidity. N Engl J Med 1985; 312:82–90.

McGauhey P, Starfield B, Alexander C, and Ensminger M. The social environment and vulnerability of low birthweight children: a socio-epidemiological perspective. Pediatrics 1991; 88:943–53.

NCHS (National Center for Health Statistics). Health, United States, 1989. Hyattsville, Md., Public Health Series, 1990, Table 21.

NCHS (National Center for Health Statistics). L Fingerhut, 1989. Trends and Current Status in Childhood Mortality, United States 1900–85. Vital and Health Statistics. Series 3 No. 26. DHHS Pub No.(PHS) 89-1410, Public Health Service. Washington, D.C., U.S. Government Printing Office.

OECD (Organization for Economic Co-operation and Development). Living Conditions in OECD Countries: A Compendium of Social Indicators. Social Policy Studies No.3, Paris, 1986.

Schroeder S. Western European responses to physician oversupply. Lessons for the United States. JAMA 1984; 252:373–84.

Starfield B and Holtzman N. A comparison of effectiveness of screening for phenylketonuria in the United States, United Kingdom and Ireland. N Engl J Med 1975; 293:118–21.

Stephen WJ. An Analysis of Primary Medical Care: An International Study. Cambridge, Cambridge University Press, 1979.

Swedish Health Service. Primary Health Care Today. Some International Comparisons. (HS 90) Stockholm, Sweden, 1981.

Townsend P. Individual or social responsibility for premature death? Current controversies in the British debate about health. Intl J Health Services 1990; 20:373–92.

Verbrugge H. The national immunization program of the Netherlands. Pediatrics 1990; 86 (6, part 2):1060–63.

Weiner J. Primary care delivery in the United States and four northwest European countries: comparing the "corporatized" with the "socialized". Milbank Mem Fund Q 1987; 65:426–61.

Whitehead M. The Health Divide: Inequalities in Health in the 1980s. London, Health Education Council, 1987.

Williams B and Miller CA. Preventive Health Care for Young Children: Findings from a Ten-Country Study and Directions for United States Policy. Arlington VA, National Center for Clinical Infant Programs, 1991.

World Bank. World Development Report 1990. Oxford, Oxford University Press, 1990.

World Health Organization. World Statistics Annual. Geneva, 1986.

16

A Research Agenda

The attainment of additional knowledge is the basis for progress in every field of endeavor. Primary care is no exception. The previous chapters of this book have documented the current state of knowledge of the organization and delivery of primary care services; the astute reader will recognize that there is still much to learn! This chapter identifies those areas that are of high priority for research in primary care.*

Two main types of research are needed to promote the effectiveness of primary care: basic research and policy-related research. Basic research includes the methods that are important in measurement, and policy-related research contributes information that helps in decision-making.

Basic Research

Many aspects of basic research would greatly benefit the measurement of primary care. This section summaries the issues of special importance in enhancing knowledge about primary care problems and primary care practice.

Measurement of Health Status. The ultimate justification for a health services intervention is its impact on health status. As noted in Chapter 2, health status has several components; any, either individually or in combination, may be influenced by health services. Although there are several techniques for assessing health status in adults (McDowell and Newell, 1987), much remains to be done, particularly in the development of measures that are clinically useful. Another prime area for developmental work concerns health status of children; the need is great for methods that are appropriate in the different age periods in childhood (Starfield, 1987).

Measurement of Case Mix and Severity of Illness. Comparisons of effectiveness of care, either over time or across facilities or health systems, should take into

*Many of the specific areas for research suggested in the following sections of this chapter are adapted from an agenda for research in primary care developed by the Agency for Health Care Policy and Research. (Nutting P. A Research Agenda for Primary Care: Summary Report of a Conference. U.S. Department of Health and Human Services Agency for Health Care Policy and Research. Rockville, Md., 1991.) Greater detail on the justification for some of these areas of needed research can be found in that document.

account initial differences in the extent or degree of illness of the different populations. Although there are several techniques for measuring this (Gold, 1988) none has been sufficiently well developed to gain widespread acceptance.

Primary care needs a case-mix measure that facilitates the categorization of individuals with more than one diagnosis or type of morbidity. One such technique has already been developed but requires wider testing in a variety of settings. (Starfield et al., 1991). Another needed development is a multiaxial coding system that would make possible the describing of co-existing health problems. Finally, there is a need for measures of severity of illness, which would make it possible to standardize or stratify populations that may differ in the severity of their illnesses (Stein et al., 1987).

Procedures for Assessing Quality of Care. Most assessments of the quality of care address only single diagnoses. Little is known about the extent to which judgments of the quality of care are consistent across diagnoses, and no techniques are available to assess this. Interventions to improve the quality of care depend on whether care is consistently poor or is poor in only a few areas; consistently inadequate care calls for a different type of intervention than does inadequate care for only one type or just a few types of specific health problems. There is a need for quality of care measures that do not depend on the presence of particular diagnoses. Such measures would be of considerable usefulness in primary care, since many of its health problems never reach the stage of a diagnosis and, conversely, patients may have several coexisting and interacting health problems under care at any one time.

The capacity to improve quality of care might be extended by developing new types of techniques. In particular, what might be gained by mechanisms that involve patients in designing and conducting quality enhancement activities? To what extent is satisfaction with care on the part of patients an adequate proxy for professional judgments about the quality of care? What particular components of satisfaction are most related to these other types of judgments about quality, and to what extent do they enhance knowledge about effective care above and beyond professional assessments?

Measurement of the Need for Referral and Characteristics of Referral Care. Which problems are best referred to specialists and which are best managed within primary care? Which types of problems call for short-term referral with the expectation that the consultant will provide advice and guidance to the primary care physician, who will then reassume care of the patient? Which type of problems call for long-term (and probably permanent) referral, with most if not all care being assumed by the tertiary care specialist? In some instances, for example, the latter might also assume primary care when the therapy as well as the management of complex malignancy can affect the health of the patient. What *other* types of problems call for such long-term care by specialist? To what extent can specialty care be streamlined so that patients who no longer need ongoing care from specialists are promptly returned to their primary care physician? Can methods be developed to distinguish consultant (short-term) care from referral (long-term) care so that specialists are appropriately trained for their respective roles, making the organization of referral care more rational?

Development and Adaptation of Methods to Manage Presenting Problems. Most assessments of the quality of diagnostic workups start with a study of the diagnosis and retrospectively examine the adequacy or appropriateness of the procedures used to reach it. With the exception of the algorithm approach to assessing quality of care, no techniques start with the presenting problem. Chapter 10, concerning medical records and information systems, summarized a method of classifying the presentation or problems in primary care. Techniques to apply it would enable facilities of health systems to judge the adequacy and appropriateness of all problems presented by patients, not only of those that resulted in particular diagnoses. Emphasis on presenting problems is even more important in primary care than in other types of care because of its first contact feature and because evidence indicates that about half of all "diagnoses" in primary care visits do not resolve into codable diagnostic entities.

What methods are best for examining how patients and practitioners negotiate their differences regarding the needs and opinions about the cause and management of health problems and determining how they arrive at a solution that leads to a mutually acceptable plan of approach?

Techniques to Measure the Effectiveness of Care and Assessment of Patients' Cooperation. This involves the degree to which patient's understand, accept, and participate in the process of care as prescribed by health professionals. Comparisons of continuous versus episodic care and evaluation of patients' participation in their care require the development of methods to collect information about patients more accurately, conveniently, and confidentially. Among the potentially useful techniques are health diaries, medical records, and abstracts of personal medical records kept by patients themselves.

What types of communication styles are most suited to patients from particular cultural subgroups? To what extent can differences across cultural subgroups be categorized and used for teaching primary care trainees?

Procedures for Assessing the Responsiveness of Patients to Medical Recommendations. There is a need for determining the point at which nonresponsiveness suggests inappropriate diagnosis or inappropriate therapy. Another area of research concerns the development of guidelines to help practitioners decide when a referral to a consultant is desirable.

Improving the Accuracy and Efficiency of Data Collection and Record Keeping. This will result in improved auditing, patient follow-up, and coordination of care. Possible approaches range from patient-carried records to electronic information networks in primary care. Another approach is basic research directed at the development of adequate information systems that would enhance the usefulness of large data bases deriving from claims records and other accounting procedures. Other techniques that might be explored include small areas surveys and repeated observations of patient cohorts to monitor care over time.

Development of a Method to Estimate the Community Served by a Primary Care Practice. The estimation of this "population at risk" is needed as a denominator for calculating rates, thereby permitting comparisons of relative frequencies among different settings or between different primary care practices.

Development of Methods to Facilitate the Use of Community Data in Primary Care Practice. Many, if not most, local and state health departments collect data that can be related to census tract characteristics. These data are important in identifying health problems that should be addressed in primary care practice. Many communities contain one or more subpopulations that are at high risk of health problems and their sequelae. These populations include the elderly (particularly those living alone), high-risk pregnant women and their infants, minority populations, socioeconomically disadvantaged populations, migrant workers, individuals with disabilities, handicaps, or impairments, and individuals at particular risk of specific diseases such as AIDS, hypertension, occupational hazards, environmental exposures, or stress-related illnesses. Examples of studies include design and development of coordinated information networks that will enable health departments, community health agencies, and primary care practitioners to collect, analyze, and interpret data on health needs. The network will also make possible the development of plans for dealing with current and emerging health problems.

Development and Testing of Ways to Examine the Effectiveness of Primary Care Training, Certification, and Educational Activities. Methods are needed to compare the various ways in which residency programs are structured, the differences and consistencies in educational approaches of the various primary care specialties, and the variations in training of physician and nonphysician practitioners. Additional methods are needed to seek out individuals who have trained in various types of programs and to assess the impact of their training on the nature, extent, and adequacy of primary care practice. Such assessment would be useful in helping training programs improve their direction and intensity, as well as in guiding their continuing education activities.

How do physicians deal with scientific information, particularly when it contradicts their current mode of practice? What characteristics determine whether physicians can alter their practices on the basis of their own reading of the evidence or whether they require more formal group mechanisms based on group processes or on the advice of respected mentors or colleagues?

Policy-Related Research

Policy-related research addresses issues that pertain to decisions about services, either directly by providing answers to important questions or indirectly by providing information that helps one understand a problem so that alternative solutions can be posed. In primary care, there are three types of policy-related research: clinical research, health services research, and research related to primary care training. Policy-relevant clinical research (sometimes known as *clinical epidemiology*) concerns the processes of care directed at recognition of symptoms, signs, and syndromes in primary care practice, the diagnostic processes associated with them, and the treatment and reassessment strategies that follow from the diagnosis. Health services research concerns the myriad relationships between the various components of the structure, process, and outcomes of care described in Chapter 2. Research related to primary care training concerns the characteristics and impact of the process to educate primary care practitioners and to maintain their competence in practice.

Clinical Primary Care Research. Within this area are several aspects in need of further investigation.

• Descriptions of the practice of primary care in various organizational configurations, settings, and communities, including the incidence and patterns of diseases seen, services provided, and resources employed. Such information will provide the basis for understanding the reasons for variations in practice patterns across different facilities. How much of the variation is a result of differences in the patients and their illnesses, and how much is a result of differences in specific mode or organization, financing, and practice? How much of the variation in use of tests, procedures, and referrals can be explained by practitioners' discomfort with uncertainty, how much to differences in patients' desires and demands, and how much to the conditions of the practice itself?

• Evaluations of the effectiveness of the drugs, devices, and procedures common in primary care practice. What are the benefits and risks of technologies or interventions employed in practice, compared with the result of testing under controlled conditions of research? What is the impact on effectiveness of various characteristics of patients, providers, and settings?

• Design and testing of protocols for screening, diagnosis, and treatment. Protocols seek to provide more systematic review of the quality and of the costs of care, to permit the identification and evaluation of alternative services or providers, and to identify problems in presenting symptoms, disease incidence, or management effectiveness. Limited studies of the use of protocols for common conditions in pediatric practice suggest that they may be of some benefit; does the use of such protocols make it possible for nonphysician providers to assume some of the burden of care in busy ambulatory facilities?

• Evaluation of prevention, patient education, and self-care activities. Although prevention of disease and promotion of health involve activities that extend beyond the health services system, aspects requiring changes in health behaviors of individuals are largely within the province of primary care. The effectiveness of various approaches, especially those based on group rather than on individual efforts, needs to be tested. Evaluation of the role of support groups in the management of common conditions such as childhood asthma or low back pain in adults would contribute in a major way to understanding the potential usefulness of group procedures rather than individual strategies to manage illness as well as to prevent it.

• Analyses of the process of medical decision-making, including both careful descriptions of how diagnostic and therapeutic choices are being made and the development and testing of methods to improve the process. How do primary care practitioners with different types and levels of training identify indicators of illness and plan management and follow-up strategies? Which prescribed strategies for decision-making are best for improving the effectiveness and efficiency of these processes, and what are their limitations and risks? Studies of this type include analyses of the role of mathematical models in medical diagnosis and of the use of utility theory in estimating and comparing preferences of patients for alternative treatments and outcomes.

• Examinations of the interrelationships between the physical and psychosocial aspects of illnesses, particularly as they are seen in primary care. Primary care is directed toward the treatment of persons rather than of diseases; practitioners frequently must sort through a wide variety of complaints and problems and assign them the correct diagnostic labels and therapeutic strategies. The interplay of physical and psychosocial disorders in patients may result in serious distortions in signs, symptoms, and treatment responses, and there is therefore a great need for information about how these interrelationships can be more easily recognized and effectively managed. Studies of this type include examinations of the relationships between stress and common complaints; analyses of the ways tranquilizers are used in office practice; and comparisons of nonpharmacologic and pharmacologic approaches to alleviating of health problems.

• Examinations of differences between users and nonusers of health services by comparing individuals, families, communities, or practices. Such factors as variations in economic status, education, cultural values, life-style, family and community support systems, and practice outreach activities may underlie the differing perceptions of providers, patients, and the public concerning necessary, appropriate, and adequate primary care. Studies of this type include examination of the adequacy and acceptability of care provided in medically underserved areas and investigation of barriers to seeking care in various populations subgroups.

• Descriptions of the natural history of illnesses commonly encountered and managed in primary care practices. More complete information is needed, for example, about the course of disorders such as otitis media, arthritis, asthma, depression, and low back pain—disorders which are characterized by cycles of remission and exacerbation. Particularly important is the extent of functional impairment over time, the effects of coexisting physical or mental disorders, and the effectiveness of efforts to prevent recurrence or exacerbation.

• Can good primary care reduce the likelihood of comorbidity in certain individuals or in particular population subgroups? For example, socioeconomically disadvantaged populations are not only more likely to become ill, they also are more likely to become ill with a variety of different types of illness (Starfield et al., 1991). Is this a result of more vulnerability to illness or a result of less adequate primary care? Can access to primary care that achieves longitudinality, comprehensiveness, and coordination reduce this tendency to comorbidity in these individuals and populations?

Health Services Research Related to Primary Care. The second area of policy-related research concerns several issues related to achieving the important functions of primary care and their interrelationships, to organizing and financing primary care, and to the challenges of increasing technology available in the office and home.

Studies related to longitudinality of care. Does a personal relationship with a particular physician better achieve the goals of primary care than a relationship with a particular place? What are the conditions under which a relationship with a particular place might achieve the same ends as a relationship with a particular doctor? If coordination of care within the different personnel in a facility is at a high level, might having a "particular place" rather than a "particular physician" be just as advantageous in achieving the benefits of longitudinality? To what extent can the benefits of identification with a particular provider be achieved by methods other than continuity of practitioner, and what are the conditions under which such alternatives are useful? What are the specific situations in which the maintenance of a long-term relationship between patient and practitioner might be detrimental to care? What problems arise when patients prefer a degree of autonomy and a less personal relationship with a provider than that which ordinarily develops in a primary care practice and which might compromise the effectiveness of care by reducing the patient's participation?

Studies related to first contact care. How can the benefits of the gatekeeper role be maintained despite incentives that may reduce access to needed specialty services? Why do some patients chose to bypass primary care physicians and go directly to specialists? Are there cultural variations in the acceptability of the primary care role? Do these differences results in different patterns and outcomes of care? What might be the disadvantages of discouraging the practice of self-referral to specialists?

Studies related to comprehensiveness. What problems are best managed solely within primary care? Conversely, what problems should routinely be referred to secondary consultants for advice or to tertiary care specialists for management? What role does increased empowerment of patients play in the improved recognition of problems by practitioners? What can be done to maximize the involvement of primary care practitioners in community outreach, long-term care, and preventive activities?

Studies related to coordination. What mechanisms can optimally facilitate the communication of information about patients between practitioners and between levels of care?

Studies concerning the relationships among the essential features of primary care. To what extent does the maintenance of a long-term practitioner–patient relationship facilitate first contact care, comprehensiveness, and coordination? Does greater comprehensiveness of care foster greater coordination? Are physicians or organizations that perform well on one characteristic likely to perform well on others, or are they unrelated? What factors are responsible for good performance on these characteristics? Are they primarily educational or are they largely a function of the setting in which the practitioner works? What are the best ways to improve performance of first contact, longitudinality, comprehensiveness, and coordination?

Studies related to referral practices. Are the observed variations in referral rates related to differences in needs of patients, or are they primarily due to either random or systematic differences in the patterns of physician practices? To what extent do differences in referral practices influence the health status of patients? Variability in referral rates might be associated with differences in (1) case mix, (2) patient characteristics such as sex and age, (3) physician characteristics such as training or duration in practice, (4) availability of resources within a practice setting as well as external ones, (5) accessibility of specialty practices to patients, and (6) differences in patterns of practice possibly related to either psychological variables associated with tolerance of uncertainty or general norms in the community. These characteristics require consideration when one explores variations in referral rates, understanding their basis and devising ways to change them to achieve the best outcomes expressed in terms of the health of patients.

Studies related to modifying pattern of referral. Studies in Great Britain show that specialists treat patients in instances when the latter could be treated by their general practitioner, and that general practitioners express dissatisfaction when patients are cross-referred to other specialists without their consent (Wilkin and Dornan, 1990, p.24). To what extent can consultants' review of their medical records remind them of the involvement of the primary care physician? What other mechanisms for involving consultants could reduce the frequency of referral? For example, can the need for referrals be reduced by holding phone conversations between the consultant and the primary care physician (Hartog, 1988) or specialty consultation sessions within the primary care setting (Tyrer, 1984)? These issues are all appropriate subjects for research.

Studies concerning the roles of primary care physicians and those of specialists. In many countries, the tasks of primary, secondary, and tertiary care physicians are clearly delineated. Primary care physicians only work outside of the hospital, consultant specialists work in community hospitals, and tertiary care physicians work only in regionalized teaching and medical centers. In some countries, secondary care physicians may work outside of hospitals as well as in them. To what extent does restriction of the practice site of primary care and specialist physicians influence the quality of care, its effectiveness, and costs?

Studies related to teamwork in primary care. Many useful contributions to primary care are made by personnel other than physicians, although there are no standards for the different roles and relationships. As noted in Chapter 9, the delegated tasks may substitute for those ordinarily performed by physicians, supplement what the physician does, or complement physicians' conventional activities. The training of nonphysician personnel for primary care would be greatly facilitated by information on what function is most appropriately assumed by whom. How widespread is the use of nonphysician personnel in providing clinical aspects of care and how variable are their tasks in primary care? Are nurse practitioners more effective than physicians in ensuring the provision of needed preventive services, and if so, are there barriers to their deployment for these purposes? How often are home visits done, for what purpose, and by whom? Under what conditions do patients relate to a nonphysician as their "regular source of care" or as the individual who coordinates the various aspects of care, and how well are the major attributes of primary care achieved compared with care by physicians?

Studies related to the organization and financing of primary care services. What financing and organizational arrangements best facilitate problem recognition, adequacy of diagnostic and therapeutic procedures, and reassessment after the institution of therapy? To what extent and under what conditions can telephone management substitute for face-to-face visits? What forms of organization and financing facilitate the development of information systems to enhance knowledge of community health needs, the recognition of patients' problems, and the responsiveness to interventions? How can we maximize the incentives to develop such systems?

Studies related to the use of technology in primary care. As technology expands, more procedures will be done at the level of primary care, and some will even be suitable for use by patients themselves. The expansion of technology beyond the confines of hospitals and large medical centers raises a host of issues for research (Young et al., 1989; Stoeckle, 1989). How does the availability of technology alter the nature of primary care? Does the availability of home testing increase or decrease the use of health services, and does it influence patients to bypass primary care physicians and seek care directly from specialists? Will new testing technologies directed at screening for early stages of disease deflect attention from the need for its prevention? Will the widespread availability of home testing impose greater burdens of coordination of services on primary care physicians? What information systems are required to facilitate the efficient use of data from these home and office based technologies? Will increasing use of technology facilitate or will it hamper the expansion of organized and integrated primary care services?

To what extent can concerted efforts to deal with problems in a family context facilitate both the process of care and its outcome? Do practitioners who use the family as an integral part of management achieve more rapid or more complete resolution of problems than primary care physicians who deal with the family context only as an adjunct to management?

Research on Primary Care Training. The third type of policy-related research has four areas that especially need investigation.

How can knowledge about the distribution of disease be included in medical curricula?

By what mechanism can primary care physicians learn the principles of primary care, and how can its content be made more adequate?

What techniques should be used in teaching physicians-in-training the limitations of their expertise as well as the proper time for appropriate consultation and referral? Since much of primary care requires dealing with uncertainty, physicians-in-training must be taught how to recognize when "watchful waiting" has reached its limits and action is required. The primary care trainee should also be taught techniques to minimize the anxiety that accompanies uncertainty. How do physicians involve patients in dealing with uncertainty? What information do they provide to patients about the likelihood of the various possibilities concerning their undifferentiated problems and the prognoses associated with them? How do these mechanisms differ with cultural background of the patient? What are the best mechanisms to teach the appropriate techniques in the educational setting?

How can approaches to quality assessment and quality improvement be incorporated into physician training? Primary care trainees should learn how to keep abreast of scientific advances in subjects related to primary care, and how to use computers to search the literature for current knowledge when it is required in the care of patients. Is such training best accomplished in conjunction with clinical activities or as part of basic science training?

Mechanisms to Facilitate the Conduct of Research in Primary Care

A very small proportion of published research concerns topics of importance in primary care. With the exception of a few journals devoted entirely to topics in family medicine and in general internal medicine, fewer than 10 percent, and in most cases well under 5 percent, of papers published in peer-reviewed journals have any relevance to the problems encountered in primary care practice (Starfield, 1990).

A major problem in primary care research concerns studies conducted in one or just a few facilities, using a population that may not be representative of the general public. About half of the research on primary care is conducted in hospital outpatient departments, with patient populations composed largely of inner-city populations in major cities. The ability to generalize findings is compromised when the study population is not representative of the population or, at least, of an identifiable population subgroup. Generalizability is a challenge in any kind of research, but it is particularly problematic in primary care because the characteristics of the particular setting are so influential that they become a constellation of factors that themselves must be considered. Therefore, a promising alternative approach to research in primary care involves building collaborative networks of research in office-based practice. Three such networks already exist: ASPN, COOP, and PROS.

Ambulatory Sentinel Practice Network (ASPN), initiated in the late 1970s by the North American Primary Care Research Group, consists largely of family physicians (Culpepper and Froom, 1988). In the mid-1980s, the American Academy of Pediatrics organized a similar network, Pediatric Research in Office Settings (PROS). Both networks involve practices across the country in conducting research on topics derived from experience in practice and are addressed at topics important to primary care practitioners. In both networks academically based physicians are of value in research management and providing the research skills necessary for the conduct of scientific research. Their steering committees also include active practitioners who participate in the choice of topics for study.

A few regional networks complement these national networks. The Cooperative Information Project (COOP) is a network of practitioners in New England who have carried out collaborative research in conjunction with Dartmouth Medical School for over a decade. Another regional network is the Pediatric Practice Research Group (PPRG) in Chicago, a joint effort of the Children's Memorial Hospital and its Institute for Education and Research, Northwestern University, together with physicians in office practice. This network has undertaken and published the results of studies on infant growth, the exposure of children to environmental hazards, the office management of foreign body ingestion, cholesterol screening, and the acceptance of new infant formulas (Christoffel et al., 1988).

PROS and its precursor based in practices in California has conducted studies of vision screening practices, iron-deficiency anemia, and fever in infants. ASPN studies have provided new information on various characteristics of a number of important problems in primary care including chest pain, pelvic inflammatory disease, spontaneous abortion, and headache (Rosser et al., 1990; Freeman et al., 1988; Green et al., 1988; Becker et al., 1988). COOP projects have addressed measurement of functional status in physicians' offices (Nelson et al., 1990), a prospective study of fatigue (Kirk et al., 1990), cancer control in primary care practice (Dietrich et al., 1990), the use of antibiotics in primary care, and many other topics (Dartmouth COOP Project Bibliography, 1991).

Over the years, many local collaborations between academics and office-based practitioners have contributed important information for primary care (Christoffel et al., 1988). Most of these collaborative efforts have been outgrowths of the interests of particular researchers based in medical centers and have therefore focused on topics generated within the centers; few efforts survived after the particular researcher moved on or found other interests. Nevertheless, the potential for broader and more stable linkages between the academic primary care physician and the office-based practitioner is evident from the existing national and regional networks.

Credible research on primary care, be it basic research, clinical research, health services research, or evaluations of education programs, requires that researchers be trained in the techniques of research. Their expertise must be specific to the task; it must include disciplines such as epidemiology, social sciences techniques, economics, and decision analysis. In addition, training for research should include techniques for collecting nonquantitative data using the skills of the anthropologist and related disciplines. Much is to be offered in the way of analyses that have become increasingly refined and are well used in conjunction with the more traditional quantitative techniques and biostatistical applications (Norton et al., 1991). For example, qualitative analyses have revealed reasons for referral that may be missed by more traditional quantitative classifications. Referral behavior may appear to be irrational and therefore difficult to make explicit by numerative categories (Wilkin and Dornan, 1990). Training in formal techniques for qualitative analysis is available for teaching as well as for use in ongoing research (Patton, 1980; Miles and Huberman, 1984).

It is accepted practice in the clinical subspecialties for trainees to undertake postdoctoral study under mentoring from senior researchers. Fellowships for postdoctoral training in health services research, including those associated specifically for research related to primary care, are available from the Agency for Health Care Policy and Research in the U.S. Department of Health and Human Services. National Research Services Awards for research on primary care issues may also be available from other agencies in the same department.

The challenge of primary care research is an exciting one. It may well be that its rewards are greater than those of other research. Primary care has the heuristic value of contributing to knowledge common to all research, but it has the added reward of contributing to research on the cutting edge of developments in health services. Biomedical research has been responsible for the enormous advances that have led to an understanding of the mechanisms of disease and for developments of technology to manage it and prolong life. Today the cutting edge lies in the organization and delivery of services to make these advances available where they are needed, in the development of ways to determine where these needs are greatest, and in focusing more attention on documenting the impact of technology and other types of services on aspects of health status other than disease processes alone. Research on primary care and translation of research findings into policy and clinical practice are essential ingredients to the achievement of the two main goals of any health services system: to optimize the health of the population by employing the most advanced state of knowledge of disease causation, illness management, and health maximization, and to minimize the systematic disparities in health status associated with differential access to the benefits of this knowledge.

References

Becker L, Iverson D, Reed F, Calonge N, Miller R, and Freeman W. Patients with new headache in primary care: a report from ASPN. J Fam Pract 1988; 27(1):41–47.

Christoffel K, Binns H, Stockman J, McGuire H, Poncher J, Unti S, Typlin B, Lasin G, Seigel W, and the Pediatric Practice Research Group. Practice-based Research: opportunities and obstacles. Pediatrics 1988; 82(Part2):399–406.

Culpepper L and Froom J. The International Primary Care Network: purpose, methods, and policies. Family Medicine 1988; 20(3):197–201.

Dartmouth COOP Project Bibliography, COOP Annual Meeting. Department of Community and Family Medicine, Dartmouth Medical School, Hanover, N.H., February 1991.

Dietrich A, O'Connor G, Keller A, Carney-Gersten P, Levy D, Nelson E, Simmons J, Barrett Jr J, and Landgraf J. Will community physicians participate in rigorous studies of cancer control? The methodology and recruitment of a randomized trial of physician practices. Prog Clin Biological Res 1990; 339:337–81.

Freeman W, Green L, and Becker L. Pelvic inflammatory disease in primary care. Fam Medicine 1988; 20(3):192–96.

Gold M. Common sense on extending DRG concepts to pay for ambulatory care. Inquiry 1988; 25:281–89.

Green L, Becker L, Freeman W, Elliott E, Iverson D, and Reed F. Spontaneous abortion in primary care. J Am Bd Fam Practice 1988; 1:15–23.

Hartog M. Medical outpatients. J Royal Coll Phys London 1988; 22:51.

Kirk J, Douglass R, Nelson E, Jaffe J, Lopez A, Ohler J, Blanchard C, Chapman R, McHugo G, and Stone K. Chief complaint of fatigue: a prospective study, J Fam Practice 1990; 30(1):33–41.

McDowell I and Newell C. Measuring Health: A Guide to Rating Scales and Questionnaires. New York, Oxford University Press, 1987.

Miles M and Huberman A. Qualitative Data Analysis: A Source Book of New Methods. Beverly Hills, Sage, 1984.

Nelson E, Landgraf J, Hays R, Wasson J, and Kirk J. The functional status of patients: How can it be measured in physicians' offices? Medical Care 1990; 28(12):1111–26.

Norton P, Stewart M, Tudiver F, Bass M, and Dunn E. Primary Care Research: Traditional and Innovative Approaches. Newbury Park, Calif., Sage Publications, 1991.

Patton M. Qualitative Evaluation Methods. Beverly Hills, Sage, 1980.

Rosser W, Henderson R, Wood M, and Green L. An exploratory report of chest pain in primary care. J Am Bd Fam Practice 1990; 3(3):143–50.

Stamps P, Kirk J, Catino D, Clark D, Nelson E, Yuskaitis A, and Arnold J. Use of antibiotics in primary care. Quality Review Bulletin 1982; (July):16–21.

Starfield B. Child health status and outcome of care: a commentary on measuring the impact of medical care on children. J Chron Dis 1987; 40(suppl):109s–15s.

Starfield, B. Commonalities in Primary Care Research, A View from Pediatrics. Invited Lecture. In: Conference on Primary Care Research: an Agenda for the 1990s. J Mayfield and M Grady (eds.). Rockville Md., U.S. Dept. of Health and Human Services, Sept. 1990.

Starfield B, Weiner J, Mumford L, and Steinwachs D. Ambulatory care groups: a categorization of diagnoses for research and management. Health Services Research 1991; 26(1):53–74.

Stein R, Perrin E, Pless IB, Gortmaker S, Perrin J, Walker DK, and Weitzman M. Severity of illness: concepts and measurement. Lancet 1987; (Dec. 26):1506–09.

Stoeckle J. Primary care and diagnostic testing outside the hospital. Intl J Technol Assess Health Care 1989; 5:21–30.

Tyrer P. Psychiatric clinics in general practice: an extension of community care. Br J Psychiatry 1984; 145:9–19.

Wilkin D and Dornan C. GP Referrals to Hospital: A Review of Research and Its Implications for Policy and Practice. Center for Primary Care Research. University of Manchester, July 1990.

Young D, Roe W, and Strauss M. Technology and primary care in the United States. Intl J Technol Assess Health Care 1989; 5:9–19.

17

A Policy Agenda and Epilogue

The Alma Ata Declaration (see Chapter 1) recognized that primary care "reflects and evolves from the economic conditions and social-cultural and political characteristics of a country and its communities . . ." (WHO, 1978). The characteristics of each primary care system are therefore unique. No two countries will ever configure their systems in identical ways, and the primary care system may even differ within countries in regions with different historical, political, cultural, and economic characteristics.

Because countries differ in these characteristics, they differ in their policy agendas. However, some generic issues must be faced by all countries, whatever the type of primary care system. This book has addressed, albeit implicitly, these issues. This chapter makes the issues explicit and brings them together. The reader should refer to the relevant chapter for clarification of the issues surrounding the policy debate.

Access to Primary Care

Unobstructed access to primary care for all who need it is a basic principle in most industrialized nations. In practicality, limitations on resources often lead to the imposition of barriers to services. Many countries have explicit policies regarding the appropriate balance between incentives and impediments to use of services. Cost considerations may make it desirable to require patients to pay some or all of the costs of primary care services. However, studies have shown that such policies reduce the number of visits and that the reduction is indiscriminate because it influences both necessary and discretionary utilization. On the other hand, some primary care systems attempt to enhance utilization by reducing or eliminating requirements for payment or paying bonuses to physicians for certain types of valued services, such as those related to prevention of illness. Any barriers to access will have a greater impact on socioeconomically disadvantaged populations and will therefore risk increasing the disparities in health status between more advantaged and less advantaged populations subgroups. The issue for policy debate concerns the extent to which constraints on resources, particularly those related to costs of care, should dictate restrictions on access and the tolerability of the consequences of these restrictions.

Quality of Care

When research demonstrates the superiority of certain practices, the policy debate focuses on the advisability of incentives for employing them or, alternatively, disincentives for using the less adequate alternatives. There are several policy options, including educational efforts of various types and intensities, peer or group pressure, denial of payment, or more drastic sanctions such as publicity about or penalties for improper actions. Although the type of option depends on the nature and severity of the contraindicated action, the policy debate will focus on the need for rewards and sanctions and the rapidity with which they are implemented after the scientific evidence of superiority is accepted.

Incentives to encourage the involvement of both practitioners and patients in developing both indicators and criteria for quality of care is another consideration for a policy agenda. There is increasing evidence of the relationship between patients' knowledge of and attitudes toward their care and its outcomes, as measured by improved well-being. Approaches that enhance the agreement between patients and practitioners concerning the goals of care would be expected to improve the overall outcomes of care. Local councils or other third-party payers that pay for care might consider incentives to practices involving patients in setting goals and objectives, with such financial incentives earmarked for improvements in a practice resulting from joint planning of patients and practitioners.

Implementation of the Cardinal Features of Primary Care

The importance of the four unique features of primary care has been documented at least moderately well if not definitively. This evidence derives from specific studies of the benefits of the features and from crossnational comparisons that demonstrate a relationship between the degree of primary care, as measured by the presence of these features, and satisfaction of the population as well as its health indicators. The policy issue is whether the evidence is strong enough to encourage better achievement of the four features and whether standards for their attainment can be instituted.

Particular areas for consideration include the role of gatekeepers, policies concerning referrals, and the expectations for case management. In policy making, the role of the gatekeeper should be defined in a standard way so as to maintain equity across sites and settings. In view of the inevitable tendency toward specialist care rather than toward primary care, policymakers might consider disincentives for specialists to see nonreferral patients. One mechanism to do this involves paying specialists a lower fee when the patient has not been referred from a primary care physician. Case management is consistent with the goals of primary care. However, when it is superimposed on a system that does not have a well-organized primary care sector, its goals and characteristics have to be clearly defined. Without such definitions by policymakers case management, like the gatekeeper concept, will defy evaluation for accountability.

More Appropriate Referrals

In policymaking, a balance between incentives and disincentives for referral should be struck. Primary care physicians should not work under conditions where they are penalized for referring patients, because penalties will inevitably lead to denial of necessary specialist services. On the other hand, referrals should not be made when the primary care physician would be expected to have the expertise necessary to solve the patient's problem. Several mechanisms might be considered to minimize unnecessary referrals and maximize legitimate ones. Periodic visiting of primary health facilities by specialists will serve at least five purposes. First, by screening patients being considered for referral, it will reduce the occurrence of referral when the primary care physician is not sure about the justification for a referral. Second, it will facilitate coordination of care by involving direct communication between the primary care physician and the specialists and, if the patient is part of the dialogue, the patient as well. Third, it will benefit the primary care physicians by providing direct education concerning subjects about which they are unsure or not knowledgeable. Fourth, it will provide the specialists with more direct information about the context of the patient's illness. And fifth, it will greatly facilitate care for patients, who do not have to spend additional time and resources making additional visits.

Other mechanisms may also improve the justifiability of referrals. Payment of specialists by salary or capitation for a defined panel should reduce incentives for visits to specialists. Another mechanism involves periodic review by primary care physicians of their own referrals. Working together, they can explore possible reasons for different referral rates among them and suggest that physicians who have persistently high rates of referral for certain types of problems undertake continuing education in that subject.

Developing a Strategy to Achieve the Derivative Features of Primary Care

Neither family centeredness nor community orientation is a focus of primary health care systems in most countries. In some countries, a family orientation is implicit as a result of the designation of family physicians as the primary care practitioners. But no country insists on accountability for family orientation. Community orientation is an ideal rather than a reality everywhere, except perhaps in a few localized areas. Although there is intuitive appeal and apparent validity of the concepts of family and community orientation, evidence of their universal practicality does not exist. Policymakers will have to decide whether to encourage research to obtain the evidence, to proceed with planning for implementation, or to disregard the concepts as low priority. The merits of the three alternative approaches deserve debate in the policy arena.

If community orientation emerges as a priority for policy, there will have to be incentives for the development of community data systems that can be used for setting priorities, planning, and evaluation. Presently existing vital statistics and reporting systems can be adapted to make the data systems more useful at the level of the communities in which the practices exist. New data systems will have to be developed to supplement information on causes of death and disease with documentation of functional status as well. This may require the conduct of periodic surveys in communities, an effort which will require additional resources. Local referenda concerning the importance accorded to such an approach will help justify the expenditure of additional funds.

Ethical Issues Concerning the Use of Resources

A major policy concern is the use of expensive technology when the procedures are clearly unjustified. Mechanisms are generally available to deny reimbursement if a policy to do so exists. When expensive procedures are scientifically justified, the concern for policymakers is the extent to which they can be justified on grounds of cost-effectiveness. Some jurisdictions have already begun developing public commissions to deliberate the costs and benefits of medical care procedures and to deny those found to have the lowest justifiability according to cost-benefit considerations. When applied to entire populations such an approach has merit. Serious ethical questions will arise, however, if the process results in a situation where certain equally needy segments of the population are denied access to services available to other segments of the population. The policy debate must consider these ethical issues as well as the more general issue of the advisability of using cost-effectiveness as a basis for making decisions about what services and procedures are to be provided or paid for by the health system.

Encouragement and Support of Research in Primary Care

Since the vast majority of health-related research is conducted by specialists in medical centers, the base of knowledge of primary care practice is underdeveloped. Policy should encourage the conduct of such research, preferably with the direct involvement of primary care physicians who have a better grasp of the nature of the problems in primary care than specialists or nonphysician researchers. Government and private research organizations should set an agenda for research and invite proposals. Professional societies should provide extra continuing education credits or present certificates of merit to physicians who become involved in such research. Collaborative efforts in which primary care physicians develop networks to define and conduct research hold particular promise because properly conducted research occurring in a wide variety of settings has greater generalizability than research on only one or a few settings. Involvement of researchers in medical centers, even if they are specialists, might also be encouraged, since exposure to primary care issues by these researchers will enhance their understanding of the challenges of primary care. However, specialists should not define the research topic nor exert primary control over the analysis or interpretation of the findings.

International Comparisons and Analyses of Differences

The tremendous variety of approaches to various aspects of primary care throughout the world provide an opportunity for each nation to identify alternatives to current modes of operation. Some of these approaches, such as limitations of specialists to hospital practice, are ingrained within systems, and some are innovative solutions to specific priorities, such as encouragement of preventive care. It is not necessary for each country to reinvent each approach; wise policy study can take advantage of knowledge deriving from already existing "demonstrations" in other countries.

Encouragement and Support of Primary Care Training

Since the natural tendency for any system is to subspecialize, special efforts have to be made to maintain a viable primary care sector. Primary care training programs will require special financial support to remain viable and attract physicians who not only are interested in it but are also of high caliber. In the face of incentives to specialize because of higher prestige and, in most places, the promise of higher incomes, trainees will gravitate toward specialization rather than primary care. The challenge to policymakers is to make primary care training and practice at least as attractive as specialty training and practice.

Building a Capacity for Accountability

When responsibility for either the financing or the provision of direct services is undertaken by government or private agencies, there must be some explicit specification of the nature and extent of services that will be undertaken. These specifications usually take the form of guidelines to be followed. These guidelines may set a minimum set of benefits to be provided and may even specify how certain services are to be delivered. For example, they may indicate that children should be seen periodically for well-child care and indicate the periodicity. Or they may require that certain types of services, such as case management services, be available to patients.

The technical capacity for accountability today extends beyond the mere specification of requirements. It is possible to set operational goals and measure their attainment. The goals may be expressed as the attainment of certain health levels, certain levels of satisfaction of the population, or set standards for quality of care.

Today, while the technical capacity for accountability of performance exists, the community of managers and practitioners is neither sufficiently aware of the capabilities nor committed to the concept of accountability to undertake the necessary activity to ensure accountability. Therefore, the challenge to policy is the adoption of a system of incentives to encourage experimentation with methods of ensuring accountability. Bonuses could be given to service organizations that set goals and measure the degree of their attainment. To encourage interest, the organizations might be permitted to choose the area in which they wish to experiment or at least to choose from a variety of options. Potential areas for increasing accountability include immunization rates of communities of children, the implementation of advisory councils of patients, evaluation of changes in health status, linkages with public health agencies that make possible the use of community health data, and involvement in research to generate new knowledge about primary care problems and their management.

Epilogue

The goals for this book will have been achieved if the reader can add other areas for research and for policy concern and action. There are many difficult challenges for primary care throughout the world. I hope this book has stimulated thinking about these challenges and has paved the way for more concerted and informed policy deliberations.

References

WHO (World Health Organization) Primary Health Care. Health for All. Series No. 1. Geneva, 1978, p.4.

Index

Team (health care), 41, 44, 78, 145–47, 243
Technology, 252
    evaluation, 180, 240
    in primary care, 244
Tertiary care, 4, 7, 9, 18, 30, 76–77, 80, 85,
    243
Time-oriented care, 4, 41–53
Tracer conditions, 180
Training programs
    content of, 92, 103, 104, 147
    evaluation of, 204–7, 239
    funding for, 246, 253
Treatment (of health problems), 14; quality of,
    178, 182

Uncertainty, in primary care, 4, 26, 136, 161,
    243, 244
Understanding (of health problems), 15
United Kingdom, 217–18, 223, 224, 225,
    226–32. *See also* Great Britain
United States, 7, 218, 221, 223, 224, 225,
    226–30, 232–34
    organization and financing of services in,
    108–29
    primary care in, 91–106
Unrepresentative samples. *See* Samples of
    patients
UPC (Usual Provider Continuity), 49
Urgency ratio, 33
Use of services. *See* Utilization

Use-disability ratio, 32
Utilization
    for first contact, 35–38
    of primary care, 203, 208–9
    of regular source of care, 48–50
    review, 173
    of services, 15, 17, 27, 28, 32–33, 35–36,
        44, 127–28, 142–45, 241
    of specialty care, 26–28, 203
    See also Visits

Visits, 137, 139
    characteristics of, 52, 96–103
    emergency, 178
    by family income, 127–28
    first contact, 35–38, 178
    patient-initiated, 35–37, 50, 187
    prevention-related, 102
    reasons for, 96–98
    specialty, 113, 115, 117
    See also Utilization
Vital statistics, 177, 192–93, 212

War on Poverty. *See* Legislation
Well-being, perceived, 15, 16
World Health Assembly, 5
World Health Organization, 63; quality of care
    program, 179–80

Years of potential life lost, 221, 227

1200 9840 60